INSULATION MATERIALS IN CONTEXT OF SUSTAINABILITY

Edited by **Amjad Almusaed**
and **Asaad Almssad**

Insulation Materials in Context of Sustainability

http://dx.doi.org/10.5772/61361
Edited by Amjad Almusaed and Asaad Almssad

Contributors

Sanjoy Debnath, Mikelis Kirpluks, Ugis Cabulis, Maria Kuranska, Aleksander Prociak, Andris Avots, David Koren, Boris Azinović, Vojko Kilar, Numan Yüksel, Nese Dikmen, Jorge Tiago Pinto, Ana Briga Sá, Sandra Pereira, Isabel Bentes, Anabela Paiva

Published by InTech

Janeza Trdine 9, 51000 Rijeka, Croatia

Publishing Process Manager Romina Rovan
Technical Editor SPi Global
Cover InTech Design team

First published August, 2016
Printed in Croatia

Additional hard copies can be obtained from orders@intechopen.com

Insulation Materials in Context of Sustainability , Edited by Amjad Almusaed and Asaad Almssad
p. cm.
Print ISBN 978-953-51-2624-9
Online ISBN 978-953-51-2625-6

For EU product safety concerns:
IN TECH d.o.o., Prolaz Marije Krucifikse Kozulić 3, 51000 Rijeka, Croatia,
info@intechopen.com or visit our website at intechopen.com.

Meet the editors

Dr. Amjad Almusaed was born on January 15, 1967. He holds a PhD degree in Architecture (Environmental Design) from "Ion Mincu" University, Bucharest, Romania. He followed a postdoctoral research in 2004 on the sustainable and bioclimatic houses, from the School of Architecture in Aarhus, Denmark. Dr. Almusaed has more than 26 years of experience in sustainability in architecture and landscape with innovative orientation. He has carried out lots of research and technical survey work and has performed several studies, in the area mentioned above. He is an active member of many international architectural associations. He published many papers, articles, researches, and books, in different languages.

Dr. Asaad Almssad was born on January 15, 1967. He holds a PhD degree in Civil Engineering from the Faculty of Civil Engineering and Agricultural Construction, Bucharest, Romania. Almssad has more than 27 years of experience in sustainable building and civil engineering. From 1989 until now, he was employed in many institutes and universities. Now, he is an associate professor in Karlstad University, Karlstad, Sweden. He has authored or coauthored more than 30 research papers and many books, in different languages.

Contents

Preface

At present, insulation matters are one of the most significant topics in building science. They offer an essential tool for creating a healthy, economical, and comfortable living environment. The main objective of this book is to present the essential role of insulation materials in general and practical use and the importance of using eco-insulation materials. The importance of this book is in subject debates, which offer a new orientation of usage of insulation materials in general and thermal insulation in particular. In this book, we try to find an area for architects, engineers, and specialists in building a domain, where innovative tools can lead specialists in building science to new and better ways of creating a comfortable living. The book is divided into two parts and six chapters.

Part I debates the importance of the use of renewable materials such as insulation materials and is composed of three chapters; the first chapter discusses the benefit of using the unusual insulation material reviewing the usage of natural and ecological materials for thermal insulation of buildings. An organic insulation material represented as a waste material, such as straw or paper, can be used for thermal and acoustic insulation, too. In this chapter the authors write about organic insulating materials. Those materials are not well known, such as sheep's wool, human hair, or reeds, where the insulation aspect of petroleum by-products is slightly lower than that of plant/agricultural materials; however, preferring the latter to the former has many positive aspects. It is a critical aspect to know that natural materials do not damage the environment by the elimination of toxic substances.

The second chapter debates the issue of using corn cob as a raw insulation material. It is an alternative procedure of an eco-friendly insulation material. Using this element as insulation material can help expanding the farm waste material in creating an efficient thermal insulation. The results shown by this study can estimate the thermal conductivity of the granulated corn cob and the granulated corn cob particleboards.

The third chapter debates a type of insulation material largely used in India and Bangladesh.

Jute-based materials have immense potential in other industrial applications as thermal insulation material. Since the jute fiber comes under this dielectric materials, products made of jute-based materials will present these properties.

Part II debates the importance of insulation materials in practical procedures. The materials presented in this part are eco-friendly materials; this part is also composed of three chapters.

The first chapter takes in evidence the seismic aspects of the application of thermal insulation boards beneath the building's foundations and the seismic role upon thermal insulation underneath the building's foundations. This chapter shows how a newly proposed structural details in building energy-efficient buildings, would, on one hand, prevent the effect of ground floor slab and thermal bridge. On the other hand, they would reduce the possible structural integrity during strong earthquakes. The authors have proposed a new solution for the foundation detail in order to reduce the seismic forces induced on the superstructure.

The second chapter proposes that the main advantage of the insulation material is its low thermal conductivity, where the key characteristics of any thermal insulation material are thermal conductivity and thermal resistance. Different bio-based raw materials, such as rapeseed oil and tall oil, could provide abundant feedstock for PU foam production. The principal aim of this chapter is to develop thermal insulation materials from renewable resources with competitive thermal insulation and fire retardancy properties and to replace halogenated FR with more sustainable solutions.

The third chapter makes a review of some commonly used methods and techniques to measure the thermal conductivity of insulation materials. The chapter suggests that the thermal conductivity of various applications can be considered as one of the most critical challenges facing thermal, mechanical, material, and civil engineers. The main target of this section is to analyze the effective thermal conductivity and the relation for the effective thermal conductivity of insulation materials. The prediction of this property has been based on experimental results, but with an analytical approach.

This book covers methods and practical applications of a scientific work oriented to experts in architecture and engineers who have an interest in this area . We hope that many experts will find this book useful not only for an interpretation of a new reading of eco-friendly insulation materials but also as a reference for how knowledge roadmaps are developed and applied. Finally, the editors would like to thank in advance the readers for comments, suggestions, advice, and guidance for this book. On this occasion, we would also like to thank Ms. Romina Rovan, InTech's Publishing Process Manager, for her sincere effort and assistance in the management of this book.

Amjad Almusaed, PhD
Department of Architecture,
School of Architecture,
Aarhus, Denmark

Asaad Almssad, PhD
Department of Engineering and Chemical Sciences,
Karlstad University,
Karlstad, Sweden

Section 1

Ecological Insulation Materials

Chapter 1

Unconventional Insulation Materials

Neşe Dikmen and Soofia Tahira Elias Ozkan

Additional information is available at the end of the chapter

http://dx.doi.org/10.5772/63311

Abstract

Materials obtained from petrochemicals (mainly polystyrene) or from natural sources processed with high-energy consumptions such as glass and rock wools are commonly utilized for the insulation of buildings. From the perspective of sustainable development, it is important to choose easily recyclable, renewable, locally available and environmentally friendly raw materials. Thermal performance of unconventional insulation materials such as pine apple leaves, wheat straw, rice straw, rice husk/hull, coconut fibre, bagasse, date palm fibre, cellulose fibre-forestry waste, corn cob and sheep wool were investigated for this study. In addition, an experiment was conducted to compare the thermal performance of different materials being used at the Eco-Center in Turkey. As a result, it can be said that the thermal conductivity of petroleum by-products (XPS, EPS, polyurethane foam) is slightly lower than that of plant/agricultural waste materials; however, preferring the latter over the former has many hidden advantages that have great long-term impacts.

Keywords: thermal insulation, thermal conductivity, conventional materials, ecological materials, waste materials

1. Introduction

From construction to demolition, the reduction of energy consumption on upcoming constructions comes as a great challenge [1]. It is becoming more and more significant to improve the energy efficiency of buildings as a large part of the world's total energy use and greenhouse gas emissions come from such constructions [2]. When comparing more energy-efficient buildings to the standard build, fossil fuels are consumed in much less quantity, therefore reducing the emissions of carbon dioxide and sulphur dioxide into the atmosphere, particularly on a micro- and meso-scale [3].

There are ways of reducing heating and cooling loads; notable among them is the proper design and selection of building envelope and its components [4]. For the implementation of the thermal protection, currently there are no better measures than the insulation of the shell of a building [5]. In addition to the overall U-value of the corresponding component including insulation, the thermal performance of building envelope is also controlled by the material's thermal properties characterized by its ability to absorb or emit solar heat [4]. A material is usually considered as a thermal insulator if its conductivity is lower than 0.07 W/mK [1].

Thermal conductivity is the primary key property of thermal building insulation material and solution, where the typical system or objective is to accomplish as low warm conductivity as could reasonably be expected [2]. Lower thermal conductivity is a result of better quality insulation properties, which also implies a higher resistance to conduction of heat through the material, creating a barrier between the surrounding environment and the samples [6]. The insulation properties of a material are typically characterized by thermal conductivity, k (N America) or λ (Europe). Thermal conductivity is expressed in units of watts per meter degree Kelvin (W/mK) and may be expressed as thermal resistance (RSI) by dividing the material thickness (m) by the thermal conductivity producing RSI (m^2K/W) [7]. Subsequently, quantitative comparisons are to be conducted between the effectiveness of different thermal insulation materials due to increased knowledge on thermal conductivity values. As a result of countless microscopic dead air-cells, which suppress convective heat transfer by preventing air from moving, thermal insulating materials begin to resist heat flow. It is the air entrapped within the insulation, which provides the thermal resistance [4].

Increased awareness towards the environment and public health is leading to an integrated evaluation of insulation materials [8]. Materials obtained from petrochemicals (mainly polystyrene) or from natural sources processed with high-energy consumptions such as glass and rock wools are commonly utilized for the insulation of buildings. However, such materials cause significant detrimental effects on the environment mainly due to the production stage including the use of non-renewable materials and fossil fuel consumption. The disposal stage also comes with unfavourable outcomes due to end of life product reuse and recycling problems. The 'sustainability' concept introduced into the building design process encouraged researchers to develop thermal and acoustic insulating materials as a result of using natural or recycled materials [1]. The energy efficiency and sustainability of buildings are currently evaluated upon many factors, not only thermal insulation thickness and heating demand, but also according to primary energy demand, CO_2 reductions and ecological properties of the building materials. Meeting these essential properties that is crucial for a holistic assessment is increasing the demand for ecological building materials, especially renewable raw material composed insulating materials. These properties are essential for a holistic assessment [9].

When recollecting on vernacular and indigenous architecture dating back thousands of years, natural insulation materials were frequently used. For example, the thatches on English country cottages or the walls and roofs in central Chile of the indigenous *Mapuche Ruca* were commonly constructed using reeds and grasses. More recent examples are included in the UNESCO Heritage side of Sewell, a mining town in Central Chile in 1905 as well as straw bale houses dating from the twentieth century in Sandhills, Nebraska, USA, where newspapers were used for the insulation of timber walls. Since the first energy crisis of the 1970s and then the pre-occupation with sustainability that has been steadily growing since the 1990s, interest has now been refocused on these materials [10].

From the perspective of sustainable development, it is important to choose easily recyclable, renewable, locally available and environmentally friendly raw materials [11]. Considering the cost-effective, bio-degradable, durable eco-friendly green building materials, global needs are to be met on thermal rehabilitation due to these interesting properties [12]. The low thermal conductivity and fibrous characteristics from the majority of organic materials have contributed to significant improvements of thermal insulation properties after incorporation in the structure of the exterior building envelope. Natural organic materials contain a higher specific heat capacity and higher moisture sensitivity, which are different physical properties when compared to ordinary silicate materials [5].

In order to find alternative sustainable building materials as well as low technology methods, which result in more affordable and more sustainable constructions, huge efforts have been applied worldwide by the research community to comply with the comfort standards required nowadays [13]. The resulting solutions can be therefore adapted by the industry, leading to a more sustainable society. The building industry is not immune to this reality [14]. Therefore, once the use of these natural materials increases, the production costs will be reduced [15].

This study consists of three steps: First, a literature survey was conducted on thermal conductivity of natural materials. Researched materials are: pine apple leaves, wheat straw, rice straw, rice husk/hull, coconut fibre, bagasse, date palm fibre, corn cob and sheep wool. Then, the thermal data collected from various ecological buildings at the Kerkenes Eco-Center in Turkey were analyzed. Finally, thermal performances of conventional and unconventional insulation materials were compared and the advantages and disadvantages of using such materials were discussed.

2. Unconventional materials

A literature review was conducted on unconventional materials: pine apple leaves, wheat straw, rice straw, rice husk/hull, coconut fibre, bagasse, date palm fibre, corn cob and sheep wool. These materials except for sheep wool are agricultural waste and generally burned after harvesting. Investigations are done on these materials for their use of thermal insulation and infill material. This part of the paper presents information and thermal conductivity values of mentioned materials.

2.1. Pineapple leaves

One of the most cultivated crops from around the world is the familiar tropical fruit, pineapple [16]. The collection and manufacturing of pineapple produces some excess residue including leaves that are currently being treated within energy plants or sometimes simply burned [1]. Burning the pineapple leaves causes environmental problems such as pollution, soil erosion and decreased soil biological activity. Therefore, industrial utilization of this material does not only prevent air pollution, which has adverse effect on air quality, human and environmental health, but is also economically profitable for agriculturists [17]. Fibres are easily extracted from the pineapple leaf as seen in **Figure 1** [18].

Figure 1. Pineapple plant and pineapple leaf fibre [18].

Research was conducted on thermal properties of pineapple leaves. Raw materials used for the study were collected from Uttaradit province located in the Northern part of Thailand. Natural rubber latex (free formaldehyde) was used as binder for the preparation of the particle boards which are 200 × 200 mm in size and 15 mm in thickness. The boards were cut into various test samples and each measurement shows the average of three different samples from three different boards. **Table 1** shows the physical properties of particle boards. Supported by the study, the use of pineapple leaves in the construction of buildings is practicable. With board densities ranging from 178–232 kg/m^3, the pre-treated natural rubber latex can be sprayed onto pineapple fibre for the manufacturing of the particle board. Fairly low readings varying between 0.043 and 0.035 W/mK were observed on the thermal conductivity of the boards. When taking into consideration the thermal conductivity and physical properties of the pineapple leaves particle board, the boards with particle binder ratios of 1:3 with density of 210 kg/m^3 are found to be promising building materials to save energy in thermal insulation applications [17].

Particle:Binder	Moisture contents (%)	Density (kg/m^3)	Water absorption (%)		Thickness swelling (%)	
			2 h	24 h	2 h	24 h
1:2	4.99	178	376	413	19	25
1:3	4.52	210	272	310	21	34
1:4	3.77	232	190	250	20	27

Table 1. The physical properties of particle boards [17].

2.2. Wheat straw

Straw, a by-product of cereal cultivation, available in large quantities at low cost from numerous countries has been one of the main materials used for the construction of green buildings throughout the world. Deriving from the cultivation of wheat, straw is commonly used for application on buildings [1]. Bales of straw are used commonly for building construction (**Figure 2**).

Figure 2. Straw bales production in Turkey [19].

Built environment professionals find straw to be an excellent construction material and recognize that some of its limitations can easily be overcome. Straw bales can be used either as load-bearing structure or as infill wall. There are different techniques used in construction for the infill wall system including post and beam structures, and more commonly practiced, beam structures and frame (truss) with straw infill (**Figure 3**) [20]. Traditionally, straw has been used for animal bedding or burned by farmers because of the storage problems [12].

Figure 3. Straw bales used as infill wall at the Kerkenes Eco-Center [19].

Reference [21] calculated thermal conductivity of straw bale samples which were 360 mm by 615 mm with a density of 60 kg/m³ and supplied by a local farmer in the United Kingdom. Hevacomp Ltd. software MAT version 16.00 were used for the calculations. Results show that the straw bales offer good insulation values of 0.067 W/mK.

A further study was conducted with the aim to research the thermal conductivity of wheat straw by reference [22]. Investigated was the barley straw grown in Southern Germany. For approximately one year, straw was packed tightly into rectangular bales measuring 50 × 40 × 80 cm³ which were stored in a barn. The density of the bales was around 70 kg/m³ and the typical straw stalk diameters were 2–4 mm. The walls of the hollow stalks had a density of approximately 300 kg/m³. Two cylindrical samples with a diameter of 28 cm were prepared from the bale in order to measure thermal conductivity in the guarded hot plate device. The samples were pressed in the measuring device (straw stalks oriented perpendicular to heat flux) until a density of around 80 kg/m³ was obtained, which is comparable with the density of the bales. The sample's final thickness was 22 mm subsequent to pressing. With the purpose to investigate heat transfer, measurements between −200 and 800°C were used on the evacuable guarded hot plate for thermal. Temperature-dependent thermal conductivity measured in the evacuated state (λ_{evac}) and the non-evacuated state (λ) can be seen in **Table 2**.

T_m (°C)	λ ($Wm^{-1} K^{-1}$)	λ_{evac} ($Wm^{-1} K^{-1}$)
20	0.0408	0.00598
50	0.0444	0.00715
75	0.0476	0.00829

Table 2. Temperature-dependent thermal conductivity measured in the evacuated state (λ_{evac}) and the non-evacuated state (λ) [22].

Measurements for straw insulation (0.041 $Wm^{-1} K^{-1}$ at 20°C) are similar to those of conventional insulation materials used in buildings when evaluating the thermal conductivities. Straw has appealing characteristics including its interesting renewable property for insulation of buildings. For the mass market, boards are required to be available in a variety of dimensions, meaning they must be developed. As thermal coupling of the fibres via gaseous conduction is already well developed at atmospheric pressures, an additional application of binders might not increase the total thermal conductivity significantly. For conventional insulation board sizes ($5 \times 50 \times 100$ cm^3 or $5 \times 62.5 \times 100$ cm^3), thin boards or elements could therefore be produced for building applications with thermal conductivities of less than 0.045 $Wm^{-1} K^{-1}$ [22]. Due to the rigidness, strength and low cost of the straw particleboards, the applications are broad [23].

2.3. Rice straw

Rice, being the world's one of the most important food grains, is produced in at least 95 countries [24]. It is the main staple food in most of the Asian countries [25], and in rice-harvesting countries large amounts of rice residues are produced annually [26] (**Figure 4**). Unsustainable use of rice straw and open burning of it in the field produces threat to environment by producing large amount of greenhouse gas emission [25]. The potential of recycling these rice residues is significant for crop production systems [26].

Figure 4. Rice straw [27].

In reference [28], a new thermal insulation material from rice straw was investigated. Material used for the study was harvested in the agricultural field of Nanjing, Jiangsu Province in China. Rice straw boards with dimension of 300 × 300 × 40 mm were prepared and thermal conductivity measurements of the test specimens were made using the Lambda 2000 measuring device. Methylene diphenyl diisocyanate (MDI) resin and acetone were used to make the boards uniform. A new thermal insulation material from rice straws with a density of 200–350 kg/m^3 and thermal conductivity of 0.051–0.053 W/mK was developed using high-frequency hot-pressing. These residues could be a great component of construction for wall or ceiling insulation to conserve energy. The thermal conductivity, boards' density and ambient temperature were found to have great correlation. Additionally, the thermal conductivity increased as the particle size decreases, and the particle moisture content did not significantly affect the thermal conductivity of the boards observed in the results. The boards with higher density had the best physical and mechanical properties. Furthermore, by decreasing the particle size in a specific range, the properties of the boards are likely to improve, although the insulation properties of the boards would be reduced.

2.4. Rice husk/hull

Rice husk/hull, which is the outer covering of a rice kernel, protects the inner ingredients from external attack by insects and bacteria [29]. Rice husk removal during rice refining creates disposal problem because this organic waste is generally burned after harvest, which causes environmental problems [30]. Thermal conductivity of rice hulls, received from a processing plant, was measured in two laboratories. Materials tested by R&D Services, Inc. (RDS) were parboiled, whereas the rice hulls tested at the Oak Ridge National Laboratory (ORNL) were not parboiled. Equipment built and in accordance with ASTM C 518 and 305 × 305 × 51 mm test frames were used for measurement of the material [31] (**Figure 5**).

Figure 5. Rice husk in test frame [31].

The densities of the specimens were 144.3, 139.4, 155.4 and 147.5 kg/m³. Experimental research conducted in two laboratories shows that thermal conductivity of rice husk ranges from 0.046 to 0.057 W/mK [31]. **Table 3** shows apparent thermal conductivity data for the material. According to reference [32] without the use of chemical binders, rice husk can be made into hard, high-density boards.

Temperature (°C)	Density (kg/m³)	ka (W/mK)	Laboratory
7.4	153.8	0.0441	ORNL
15.5	153.8	0.0452	ORNL
23.9	153.8	0.0464	ORNL
32.2	153.8	0.0476	ORNL
40.6	153.8	0.0484	ORNL
23.9	153.8	0.0462	ORNL
7.3	168.2	0.0488	ORNL
15.6	168.2	0.0510	ORNL
23.9	168.2	0.0532	ORNL
32.2	168.2	0.0552	ORNL
40.6	168.2	0.0561	ORNL
23.9	168.2	0.0496	ORNL
23.9	144.3	0.0566	RDS
23.9	139.4	0.0477	RDS
23.9	155.4	0.0493	RDS
23.9	147.5	0.0490	RDS

Table 3. Apparent thermal conductivity data for rice husk from two laboratories [31].

2.5. Coconut fibre

Coconuts are growing abundantly in coastal areas of tropical countries [33]. From the outer shell of a coconut, coconut fibre is extracted. There are just two types of coconut fibres which are brown fibres extracted from matured coconuts and from immature coconuts white fibres are extracted. While brown fibres are strong, thick and have high abrasion resistance, white fibres are not only smoother and finer but also weaker. Coconut fibres are commercially accessible in three forms, namely bristle (long fibres), mattress (relatively short) and decorticated (mixed fibres). Depending on the requirement, these different types of fibres have different uses. Taking engineering into consideration, the brown fibres are more commonly applied [34]. Coconut tree, coconut and coconut fibres are seen in **Figure 6**.

Figure 6. Coconut tree, coconut and coconut fibres [35].

The major composition of coconut fibres contains cellulose, hemi-cellulose and lignin, affecting the different properties of the coconut fibres. The composition ultimately changes not only its properties but also the properties of composites when the fibres are pre-treated. From time to time, the fibres' behaviour improves although sometimes the effect is not favourable. The most ductile material amongst all natural fibres is the coconut fibre [35]. Since coconut fibre is a natural material, it decomposes faster and more easily. Therefore, it pertains a clean built environment. This material can be recycled and introduced as an insulation material [36].

In reference [37], the thermal conductivity of coconut fibre was investigated and a minimum material thermal conductivity of about 0.05 W/mK was found. Coconuts used for the study were obtained from coastal areas in Mexico. Horizontal fibre and vertical fibre cylindrical specimens 6 mm in thickness and 15 mm in diameter with 174 kg/m^3 density were prepared (**Figure 7**). Fibres were glued together only by compression; artificial binders were not used. A locally designed constant-temperature, one-piece specimen apparatus was used and recommendation of the ASTM norm C518 was followed for the thermal conductivity tests of the research. According to reference [32] without the use of binders, coconut husk can be made into thermal insulations.

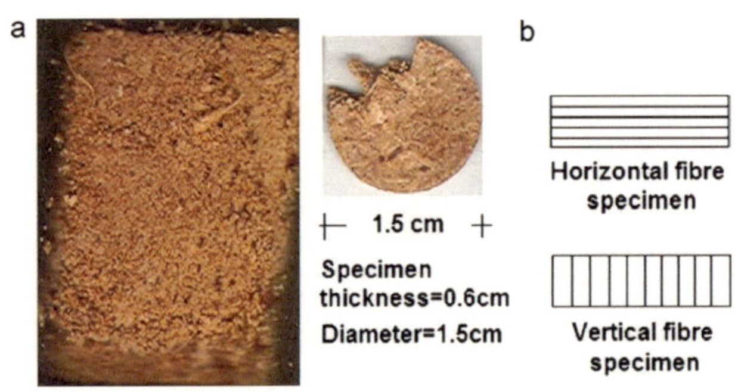

Figure 7. Material and specimen prepared for the study [37].

http://dx.doi.org/10.5772/63311

2.6. Bagasse

Sugarcane bagasse is typically found in tropical countries that process sugarcane such as Brazil, India, Cuba, Iran [38] and Pakistan (**Figure 8**). Traditionally, bagasse is managed as waste; it is burned or used as animal fodder. According to reference [1] besides its great availability, its low cost and cellulose content that helps to reduce the use of synthetic binders fostered several researchers to work towards developing innovative thermal insulation particle boards made from such material [1]. Reference [39] conducted a research on thermal insulation boards made from bagasse. Bagasse was obtained from the waste of the sugar factory in Ratchaburi province of Bangkok, and 25-mm thick test boards were produced at the target board densities of 250, 350 and 450 kg/m^3 in a laboratory at the Royal Department of Forestry, Bangkok, Thailand. Bagasse particles were formed manually using a forming box into a mat of size 450 × 450 mm and 81 binderless insulation boards were made by corresponding to three boards for each of the 27 manufacturing conditions: combinations of board density (three levels: 250, 350 and 450 kg/m^3), hot-pressing temperature (three levels: 160, 180, 200°C) and hot-pressing duration (three levels: 7, 10 and 13 min). All boards were cut and trimmed into various test specimens and thermal conductivity was measured at room temperature using a Heat Flow Meter under steady-state, one-dimensional test condition with upward heat flow. According to the results, the thermal conductivity values of bagasse range from 0.046 to 0.068 W/mK depending on the density of the material. Reference [32] states that hard (high density) boards can be transformed from bagasse without the use of chemical binders.

Figure 8. Sugarcane and sugarcane bagasse [1].

2.7. Date palm fibre

The date palm is cultivated in many regions of the world, especially the arid areas [40]. The residues of the date palm, such as leaves, petioles and bunches, are commonly considered as waste [1]. With four types of fibre, date palms consist of a fibrous structure: leaf fibres in the peduncle, baste fibres in the stem, wood fibres in the trunk and surface fibres located in the trunk [41]. Date palm tree and date palm tree fibres are seen in **Figure 9**. An experimental study was conducted on thermal conductivity of date palm fibre by reference [42]. Natural fibres, which were collected from the region of Errachidia in Morocco, were dried in oven at 60°C after washing. The thermal conductivity of the natural fibres, which were collected, are achieved according to NF ISO 88941 2nd edition 15/05/2010 using the CT meter device. The date palm fibres were separated into single fibres and measurements were done at 25°C. Results of the study show that the lowest thermal conductivity of date palm fibre is 0.041 W/mK.

Figure 9. Date palm tree and date palm fibres [43].

2.8. Corn cob

By processing corn plants the residuals of corn cobs are produced [1]. Corn cob has an advantage when thought in terms of possible application for alternative processed products because it does not collide with the worldwide food stock and it is generally considered as agricultural waste [14]. In terms of shape, texture, density and colour, the corn cob presents three layers which are much different (**Figure 10**). In contrast to common thermal insulation building materials, the corn cob material is heterogeneous which is related to its natural biological origin [44].

Figure 10. Corn cob and three layers of corn cob [44].

An experimental research on thermal conductivity of corn cob was conducted by reference [13]. Average density of the randomly picked up corn cob samples was measured as 212.11 kg/m^3. A corn cob particle board with dimensions of 250 × 250 × 50 mm was prepared and wood glue was used as binder. In addition, an XPS panel with dimensions 640 × 760 × 50 mm was used for the study. The panel system of XPS frame and corn cob particle board was replaced by a window of a confined room. The temperature in the room was kept nearly constant at 23°C. Two heat fluxmeters and two temperature sensors were placed on the inner surface of the particle board. Two thermohygrometers were placed inside and outside of the room in order to measure indoor and outdoor temperatures. Two heat fluxmeters were used to measure the heat flow through the corn cob particle board. The measurement was done continuously (10 min timing interval) for 7 days. Results of the study show that the thermal conductivity of the corn cob is 0.139 W/mK.

Reference [44] analyzed samples of corn's cob and XPS by SEM/EDS in order to compare their microstructure and elementary chemical composition. This study was conducted at the Electronic Microscopy Laboratory of the Tras-os-Montes Alto Douro University. Some interesting similarities between the corn's cob and the extruded polystyrene (XPS) materials are indicated in SEM/ED's results. The closed cellular microstructure type explains these similarities, as well as the presence of same chemical elements. The authors suggest using corn cob as a filling material and thermal insulation material.

2.9. Sheep wool

Sheep's wool is a sustainable, natural resource whose inherent characteristics make it attractive as an insulating material [8]. The shearing of sheep grow wool is a pain-free technique commonly carried out at least once a year to relieve the sheep of stress and discomfort, especially in hot and humid conditions. As a result of this non-invasive technique, sheep wool has traditionally been used for the manufacturing of conventional woollen products in the textile industry, such as carpets, garments, curtains, covers and bedding [11]. It is an easily renewable, easily recyclable and environmentally friendly source of raw material (**Figure 11**) [9].

Figure 11. Sheep wool and its application in building construction [9].

Thermal conductivity of sheep wool was found out at steady state using the plate method measured by the Lambda 2300 device, Micromet Inc., Holometrix, USA. The 300 × 300 mm samples with varying thicknesses of 80, 70, 60, 50 and 40 mm were used for the study. The samples were prepared from the manufactured sheep wool-based mats. Thermal conductivity of the samples was measured at mean temperatures +10, +20, +30 and +40°C. Depending on the density of the material, the thermal conductivity of sheep wool is 0.034–0.050 W/mK [9]. **Table 4** shows thermal conductivity and bulk density of the material. Reference [7] states that sheep's wool does have the potential to be developed into a sustainable, natural and renewable insulation material, but one that could perhaps serve local, regional or niche markets.

Sample	Thickness (mm)	Bulk density (kgm^{-3})	Thermal conductivity ($Wm^{-1} K^{-1}$)	Mean temperature (°C)
1	40	40	0.034	
2	50	32	0.035	
3	60	27	0.037	10
4	70	23	0.038	
5	80	20	0.040	
1	40	40	0.036	
2	50	32	0.038	
3	60	27	0.040	20
4	70	23	0.042	
5	80	20	0.044	
1	40	40	0.038	
2	50	32	0.040	
3	60	27	0.042	30
4	70	23	0.045	
5	80	20	0.048	
1	40	40	0.039	
2	50	32	0.041	
3	60	27	0.043	40
4	70	23	0.046	
5	80	20	0.050	

Table 4. Overview of thermal conductivity and bulk density [9].

3. Thermal performance of natural and fabricated materials

Over the span of a few years, nine buildings have been constructed with different materials and configurations at the Kerkenes Eco-Center, which is located in a village in Turkey. The aim of this ongoing applied research was to understand indigenous materials and techniques and compare them with conventional ones to see which performs better, especially from the point of view of thermal comfort. In this regard, an experiment was conducted to compare the thermal performance of different materials being used at the Eco-Center. The data have been plotted in a chart which is presented in **Figure 12**.

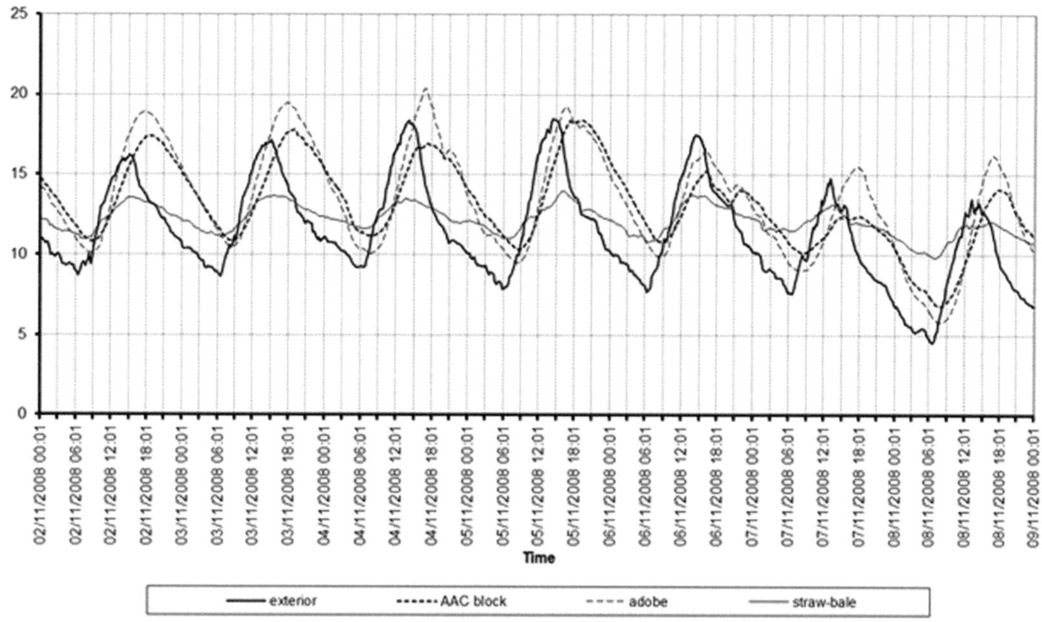

Figure 12. Comparison of materials used in Kerkenes Eco-Center in Turkey.

This study was followed by comparing the thermal behaviour of buildings located at the Eco-Center, constructed with the very same materials. Again data loggers were used to gather temperature and humidity data from these buildings at 15 min intervals and then plotted on a chart for comparison.

The chart (**Figure 13**) is a good example of the behaviour of these materials. Even with a difference in external temperature superior to 10°C between day and night, all of these buildings stayed more or less stable; the diurnal temperature fluctuations were within 2°C. From these results, we can see that buildings made of locally produced indigenous blocks or straw bales are just as thermally comfortable as those produced using AAC blocks, which is a popular construction material in Turkey.

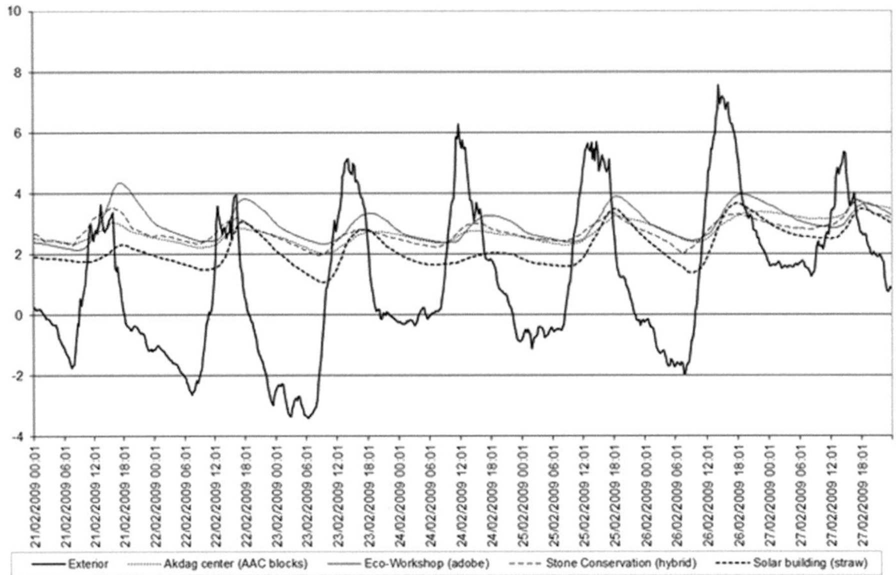

Figure 13. Comparison of thermal performance of the buildings in Kerkenes Eco-Center.

4. Conclusions

The environmentally friendly natural materials presented in this chapter are mostly considered as waste, with the exception of sheep's wool, and are usually burnt on the fields or as fuel in stoves. This method of disposal leads to an increase of CO_2 and ashes in the environment. On the other hand, the thermal conductivity value of each of these materials can be considered as comparable to that of conventional and popular insulation materials being used by the construction industry worldwide. **Table 5** shows the data collected through the survey. The table includes thermal conductivity of natural materials as well as of conventional insulation materials.

From **Table 5**, one can argue that the thermal conductivity of petroleum by-products (XPS, EPS, polyurethane foam) is slightly lower than that of plant/agricultural waste materials; however, preferring the latter over the former has many hidden advantages that have great long-term impacts. Firstly, producing natural materials does not harm the environment in the way toxic materials that are used to produce insulation boards do. Secondly, when these materials are put back into the production cycle as raw material, instead of treating them as waste to be disposed of by burning, the embodied carbon and nitrogen are prevented from being released into the environment as harmful gases and ashes. Thirdly, using these natural materials does not pose any threats to human or environmental health; and finally, the embodied energy and life-cycle costs of natural materials are considerably lower than those of toxic and environmentally harmful conventional insulation materials.

Material	Regions where these materials are mostly produced	Characteristics	Uses	Information related to the experimental studies reviewed			
				Test specimen	Binder	Density	Thermal Conductivity
Non-conventional insulation materials (thermal conductivity range = 0.034–0.067 W/mK)							
Pineapple leaves	Tropical regions	Plant waste	Particle board	Particle board	Natural rubber latex	178–232 kg/m^3	0.035–0.043 to W/mK
Wheat straw	All over the world	Agricultural waste	Straw bale/ animal fodder/ particleboard/ fibre fill	Straw bale	No binder	60 kg/m^3	0.067 W/mK
				Cylindrical samples	No binder	80 kg/m^3	0.041 W/mK
Rice straw	All over the world	Agricultural waste	Straw bale/ board/fibre fill	Board	Methylene diphenyl diisocyanate and acetone	200–350 kg/m^3	0.051–0.053 W/mK
Rice husk/ hull	All over the world	Plant waste	Board/particle board/fibre fill	Board	No binder	144.3–147.5 kg/m^3	0.046–0.057 W/mK
Coconut fibre	Coastal areas of tropical countries	Plant waste	Particle board/ fibre fill	Particle board	No binder	174 kg/m^3	0.05 W/mK
Bagasse	Tropical regions	Plant waste	Particle board/ fibre fill	Particle board	No binder	250, 350, 450 kg/m^3	0.046–0.068 W/mK
Date palm fibre	Arid areas	Plant waste	Fibre board/fibre fill	Fibre board	No binder	Not mentioned	0.041 W/mK
Corn cob	All over the world	Agricultural waste	Particle board/ fibre fill	Particle board	Wood glue	212.11 kg/m^3 (average)	0.139 W/mK
Sheep wool	All over the world	Animal fibre	Textile and clothing/mat/ fibre fill	Mat	No binder	20–40 kg/m^3	0.034–0.050 W/mK
Conventional insulation materials (thermal conductivity range = 0.029–0.045 W/mK)							
EPS			Insulation boards				0.03–0.033 W/mK
XPS			Insulation boards				0.029–0.039 W/mK
Foam plastic			In-situ insulation				0.03 W/mK
Glass-wool			Insulation boards				0.04 W/mK
Rock-wool			Insulation boards				0.045 W/mK

Table 5. Data collected through the survey including thermal conductivity of natural materials as well as of conventional insulation materials.

Author details

Neşe Dikmen[1*] and Soofia Tahira Elias Ozkan[2]

*Address all correspondence to: nesedikmen@sdu.edu.tr

1 Suleyman Demirel University, Faculty of Technology, Department of Civil Engineering, Isparta, Turkey

2 Middle East Technical University, Faculty of Architecture, Department of Architecture, Ankara, Turkey

References

[1] Asdrubali F, D'Alessandro F and Schiavoni S: A review of unconventional sustainable building insulation materials. Sustainable Materials and Technologies. 2015;4:1–17.

[2] Jelle BP: Traditional, state-of-the-art and future thermal building insulation materials and solutions: properties, requirements and possibilities. Energy and Buildings. 2011;43:2549–2563.

[3] Papadopoulos AM and Giama E: Environmental performance evaluation of thermal insulation materials and its impact on the building. Building and Environment. 2007;42:2178–2187.

[4] Al-Homoud MS: Performance characteristics and practical applications of common building thermal insulation materials. Building and Environment. 2005;40:353–366.

[5] Korjenic A, Petranek V, Zach J and Hroudova J: Development and performance evaluation of natural thermal-insulation materials composed of renewable resources. Energy and Buildings. 2011;43:2518–2523.

[6] Patnaika A, Mvubua M, Muniyasamya S, Botha A, Anandjiwalaa RD: Thermal and sound insulation materials from waste wool and recycled polyester fibers and their biodegradation studies. Energy and Buildings. 2015;92:161–169.

[7] Corscadden KW, Biggs JN and Stiles DK: Sheep's wool insulation: a sustainable alternative use for a renewable resource? Resources, Conservation and Recycling. 2014;86:9–15.

[8] Papadopoulos AM: State of the art in thermal insulation materials and aims for future developments. Energy and Buildings. 2005;37:77–86.

[9] Zacha J, Korjenic A, Petraneka V, Hroudovaa J and Bednarb T: Performance evaluation and research of alternative thermal insulations based on sheep wool. Energy and Buildings. 2012;49:246–253.

[10] Petit-Breuilh X, Whitman CJ, Lagos C, Armijo PG and Schiappacasse LN: Natural fibre insulation in rural southern Chile. In: Proceedings of the PLEA2013: 29th Conference, Sustainable Architecture for a Renewable Future; 10–12 September 2013; Munich, Germany.

[11] Korjenic A, Klarić S, Hadžić A and Korjenic S: Sheep wool as a construction material for energy efficiency improvement. Energies. 2015;8:5765–5781. doi:10.3390/en8065765

[12] Chetehouna K, Belayachi N, Rengel B, Hoxha D and Gillard P: Investigation on the thermal degradation and kinetic parameters of innovative insulation materials using TGA-MS. Applied Thermal Engineering. 2015;81:177–184.

[13] Pinto J, Cruz D, Paiva A, Pereira S, Tavares P, Fernandes L and Varum H: Characterization of corn cob as a possible raw building material. Construction and Building Materials. 2012;34:28–33.

[14] Paivaa A, Pereiraa S, Sáa A, Cruza D, Varum H and Pinto J: A contribution to the thermal insulation performance characterization of corn cob particleboards. Energy and Buildings. 2012;45:274–279.

[15] Lertwattanaruk P and Suntijitto A: Properties of natural fiber cement materials containing coconut coir and oil palm fibers for residential building applications. Construction and Building Materials. 2015;94:664–669.

[16] Yusofa Y, Yahyaa SA, Adama A: Novel technology for sustainable pineapple leaf fibers productions. In: 12th Global Conference on Sustainable Manufacturing Procedia CIRP; 2015; vol. 26, pp. 756–760.

[17] Tangjuank S: Thermal insulation and physical properties of particleboards from pineapple leaves. International Journal of Physical Sciences. 2011;6(19):4528–4532.

[18] Asim M, Abdan K, Jawaid M, Nasir M, Dashtizadeh Z, Ishak MR and Hoque ME: A review on pineapple leaves fibre and its composites. Hindawi Publishing Corporation International Journal of Polymer. 2015; Article ID 950567.

[19] Straw bale production in Turkey. Available from http://www.kerkenes.metu.edu.tr/ [Accessed: 26-02-2016].

[20] Ashoura T, Georg H and Wuc W: Performance of straw bale wall: a case of study. Energy and Buildings. 2011;43:1960–1967.

[21] Goodhew S and Griffiths R: Sustainable earth walls to meet the building regulation. Energy and Buildings. 2005;37(5):451–459.

[22] Beck A, Heinemann U, Reidinger M and Fricke J: Thermal transport in straw insulation. Journal of Thermal Environment Building Science. 2004;27(3):227–234.

[23] Mo X, Cheng E, Wang D and Sun XS: Physical properties of medium-density wheat straw particleboard using different adhesives. Industrial Crops and Products. 2003;18:47–53.

[24] Coats B. Global rice production. In: Smith CW, Dilday RH, editors. Rice Origin, History, Technology and Production: Texas A&M University; 2003. pp. 247–470.

[25] Singh R, Srivastava M and Shukla A: Environmental sustainability of bioethanol production from rice straw in India: a review. Renewable and Sustainable Energy Reviews. 2016;54:202–216.

[26] Mandal K G, Misra A K, Hati K M, Bandyopadhyay K K, Ghosh P K, Mohanty M: Rice residue– management options and effects on soil properties and crop productivity. Food, Agriculture – Environment. 2004; 2(1): 224–231.

[27] Rice straw. Available from http://naturalhomes.org/bamboo-rice.htm [Accessed: 28-01-2015].

[28] Wei K, Lv C, Chen M, Zhou X, Dai Z and Shen D: Development and performance evaluation of a new thermal insulation material from rice straw using high frequency hot-pressing. Energy and Buildings. 2015;87:116–122.

[29] Jung DS, Ryou MH, Sung YJ, Park SB and Choi W: Recycling rice husks for high-capacity lithium battery anodes. PNAS. 2013;110(30):12229–12234.

[30] Kumar A, Mohanta K, Kumar D and Parkash O: Properties and industrial applications of rice husk: a review. International Journal of Emerging Technology and Advanced Engineering. 2012;2(10):86–90.

[31] Yarbrough DW, Wilkes KE, Olivier PA, Graves RS and Vohra A: Apparent thermal conductivity data and related information for rice hulls and crushed pecan shells. Thermal Conductivity. 2005;27:222–230.

[32] Panyakaew S and Fotios S: Agricultural waste as thermal insulation for dwellings in Thailand: preliminary results. In Proceedings of PLEA 2008: 25th Conference on Passive and Low Energy Architecture; 22–24 October 2008; Dublin, Ireland.

[33] Van Dam JEG, Van Den Oever MJA, Teunissen W, Keijsers ERP and Peralta AG: Process for production of high density/high performance binderless boards from whole coconut husk. Industrial Crops and Products. 2004;19:207–216.

[34] Ali M, Liu A, Sou H and Chouw N: Mechanical and dynamic properties of coconut fibre reinforced concrete. Construction and Building Materials. 2012;30:814–825.

[35] Ali M: Coconut fibre-a versatile material and its applications in engineering. In: Proceedings of Second International Conference on Sustainable Construction Materials and Technologies; 28–30 June 2010; Ancona, Italy.

[36] Mintorogoa DS, Widigdoa WK and Juniwatia A: Application of coconut fibres as outer eco-insulation to control solar heat radiation on horizontal concrete slab rooftop. Procedia Engineering. 2015;125:765–772.

[37] Rodríguez NJ, Yánez-Limón M, Gutiérrez-Miceli FA, Gomez-Guzmanc O, Matadamas-Ortiz TP, Lagunez-Rivera L and Vazquez Feijoo JA: Assessment of coconut fibre

insulation characteristics and its use to modulate temperatures in concrete slabs with the aid of a finite element methodology. Energy and Buildings. 2011;43:1264–1272.

[38] Samariha A and Khakifirooz: NSSC for bagasse. BioResources. 2011;6(3):3313–3323.

[39] Panyakaew S and Fotios S: New thermal insulation boards made from coconut husk and bagasse. Energy and Buildings. 2011;43:1732–1739.

[40] Taallah B and Guettala A: The mechanical and physical properties of compressed earth block stabilized with lime and filled with untreated and alkali-treated date palm fibers. Construction and Building Materials. 2016;104:52–62.

[41] Kriker A, Debicki G, Bali A, Khenfer MM and Chabannet M: Mechanical properties of date palm fibres and concrete reinforced with date palm fibre in hot-dry climate. Cement & Concrete Composites. 2005;27:554–564.

[42] Oushabi A, Sair S, Abboud Y, Tanane O and EL Bouari A: Natural thermal-insulation materials composed of renewable resources: characterization of local date palm fibers (LDPF). Journal of Materials and Environmental Science. 2015;6(12):3395–3402.

[43] Alawar A, Hamed AM and Al-Kaabi K: Characterization of treated date palm tree fiber as composite reinforcement. Composites: Part B. 2009;40:601–606.

[44] Pinto J, Paiva A, Varum H, Costa A, Cruz D, Pereira S, Fernandes L, Tavares P and Agarwal J: Corn's cob as a potential ecological thermal insulation material. Energy and Buildings. 2011;43:1985–1990.

Chapter 2

Possible Applications of Corncob as a Raw Insulation Material

Jorge Pinto, Ana Briga Sá, Sandra Pereira,
Isabel Bentes and Anabela Paiva

Additional information is available at the end of the chapter

http://dx.doi.org/10.5772/62339

Abstract

Some alternative applications of corncob as a raw thermal insulation material are presented in this research work. Usually, corncob has been treated as an agricultural waste. Finding practical applications of this waste in product manufacturing may preserve the environment and may also allow using more green technologies. Therefore, a corncob particleboard, a lightweight concrete for nonstructural purposes, and a lightweight concrete masonry unit (CMU) are the granulated corncob-based products proposed. These products are studied in terms of thermal performance, and some thermal parameters are delivered. The results obtained through the experimental study allowed to estimate the thermal conductivity of the granulated corncob and of the granulated corncob particleboards. The values obtained were 0.058 and 0.101 W/m°C, respectively. A thermal transmission coefficient of 1.99 W/m^2°C was obtained for the nonstructural corncob lightweight concrete, and it was concluded that the density and the thermal properties of this alternative solution are in accordance with the properties of the currently used expanded clay concrete. For the granulated corncob lightweight CMU, a value of 1.15 W/m^2°C was estimated. This shows that this agricultural waste may have potential as a thermal insulation product.

Keywords: agricultural waste, corncob, lightweight aggregate, raw building materials, sustainability, thermal insulation material

1. Introduction

Sustainable issues are ruling modern society. Development is supported in sustainable parameters. Cities are becoming smart and green. Finding alternative sustainable building materials and low technological building methodologies are solutions to give a contribution in this context.

Sustainable and affordable construction, complemented with the comfort standards required nowadays, may be an objective to achieve in the building industry. CO_2 emissions to the atmosphere, energy and water consumptions, and affordability are some parameters to take into consideration in the perspective of green product manufacturing processes. In addition, there are other aspects that contribute to green building solutions or practices, such as reusing, opting for green building materials (which must be renewable, local, and abundant), retrofitting, and choosing low-technology methods and techniques.

Therefore, in the building industry, a range of several different products or building solutions based on the application of raw organic materials have already been experimented. Among these organic-based building materials, wood and wood-engineered products, bamboo, and cork and cork-engineered products are perhaps the most commonly applied ones. However, different agricultural products have also been reported as possible raw organic building materials [1–5], such as bagasse, cereal, straw, corn stalk, corncob, cotton stalks, kenaf, rice husks, rice, straw, sunflower hulls and stalks, banana stalks, coconut coir, bamboo, durian peel, and palm leaf oil, among others. These raw organic building materials are recommended for the manufacturing of different thermal insulation products [6].

Among the above-identified agricultural products, corncob has an advantage of being an agricultural waste [6]. In fact, the corn production amount has shown an increasing tendency during the last years; therefore, the production of corncob has also shown the same tendency [7]. The corn plant, *Zea mays*, was introduced in Portugal in the mid-16th century, and since then, it has been part of the Portuguese agricultural sector. As far as known, there is still no practical application of this organic product [6]. In other countries, such as the United States, China, and Brazil, the plantation of corn is very relevant in the agriculture context and the corn production per year has been increasing in these countries. Taking into account this context, finding innovative applications for this agriculture waste may result in an alternative and sustainable product that may be relevant, taking into account the overall amount of corncob produced worldwide per year [6].

There are several research works [8–14] that have given particular emphasis to the application of corncob product processing.

This chapter intends to present three different building products based on granulated corncob, such as a particleboard (case study I), a lightweight concrete for nonstructural purposes (case study II), and a lightweight concrete masonry unit (CMU; case study III). These products are described in brief, and their thermal insulation behavior is also studied experimentally.

Thus, apart from this introduction section, this chapter is structured as follows. First, an example of a traditional building application of corncob as a filling material of walls is introduced. Second, some material properties of corncob (such as macrostructure, microstructure, density, water absorption, and fire resistance) are delivered; also, a brief comparison is done with current applied thermal insulation materials. Third, the adopted experimental thermal insulation performance evaluation is described in which the experimental set-up, the facility, the equipment, and the calculation methodology are included. Fourth, the main results are presented and discussed. These results are concerned with the three corncob-based products (case studies I–III). Finally, some remarks are done in the conclusion section.

2. An example of a traditional building application of corncob as a filling material of walls

To exemplify that corncob may be already a building material, **Figure 1** shows an ancient *tabique* wall that presents corncob as a filling material, which also was already reported in [15].

Figure 1. A traditional external *tabique* wall showing corncob as a filling material.

It is worth to explain that *tabique* construction is one of the main Portuguese traditional building techniques that use earth-based building materials. In fact, a common *tabique* building element (such as a wall; **Figure 1**) is usually formed by a regular timber structural frame (made of vertical boards and laths), which is filled and covered by earth or an earth-based mortar. In certain cases, organic filling materials, such as wood pieces, straw, or corncob, are also used and mixed up with earth. In the case of the *tabique* wall presented in **Figure 1**, earth has already been removed, and corncobs and the timber frame are completely exposed.

3. Some material properties of corncob

Corncob is a raw organic material, and therefore a relevant material heterogeneity is expected. Material discontinuity and anisotropy are also two material characteristics of corncob. In terms of macrostructure, corncob tends to show three distinct layers (layers I–III, from inside to outside; **Figure 2**), which are clearly perceived by their colour, texture, shape, and density [15]. Layer I is quite soft, layer II is similar to solid softwood, and layer II is very irregular. **Figure 3** shows the microstructure of layer I of a corn cob, which is a closed cellular structure (alveolar) is similar to a typical thermal insulation material such as extruded polystyrene (XPS) or expanded polystyrene (EPS).

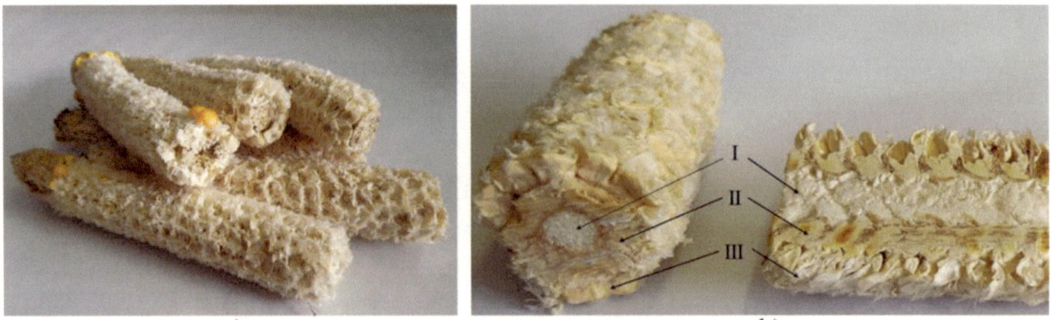

a) b)

Figure 2. Corncob macrostructure: (a) general view and (b) longitudinal section [15].

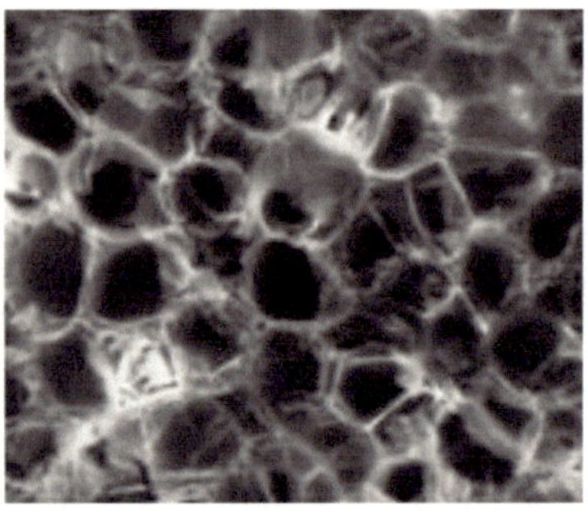

Figure 3. Typical microstructure of layer I of corncob (400 μm).

Meanwhile, the density of corncob was attempted to be experimentally measured [6]. In that study, the average value of the measured density of corncob was 212.11 kg/m^3, with a coefficient of variation of 22.4%, which was expectable considering the material heterogeneity described above. Comparing this evaluated density with the respective property of currently applied thermal insulation materials, it is concluded that corncob density is significantly higher than the densities of the XPS and EPS, which are 25–40 and 10–25 kg/m^3, respectively [16]. The densities of corncob and cork may be considered quite similar, taking into account that the density of the cork varies between 100 and 350 kg/m^3. On the other hand, and in this respect, expanded clay presents higher density values, which may be in between 275 and 430 kg/m^3 [16]. It is worth to remind that most of these materials are processed (i.e. XPS, EPS, and expanded clay). Only corncob and cork are natural and organic materials, and they seem to have similar densities.

In terms of water absorption, corncob seems to have an impressive water absorption capacity (327%), which contrasts with XPS, EPS, and expanded clay that present the approximate water absorption value of 13%, 34%, and 36%, respectively. On the other hand, granulated cork presents a value of water absorption of 244%, which is more similar to the one of corncob. This fact converges to the assumption that corncob and cork may have interesting material property similarities. It was noticed that corncob has a progressive saturating process, which may be related to the microstructure described above [6].

Fire resistance may be other material property that is convenient to assess for a building material and, therefore, for corncob. According to Santos [16], an expedite fire resistance test was performed. Flaming, combustion, gas emission, and the time consumption for total combustion were the data collected during this test. Samples of corncob, XPS, EPS, cork, and expanded clay were exposed to a direct flame for 5 minutes, which was the maximum test duration considered. Expanded clay was not affected by the direct exposure to a flame. On the other hand, corncob and granulated cork showed similarities in what concerns to the fire resistance behavior. In this case, both organic materials showed a slow progressive combustion process characterized by flame and a black gas emission. XPS and EPS were the most vulnerable insulation materials under fire exposure. Therefore, corncob seems to have an acceptable fire resistance when compared with current applied thermal insulation materials [6].

4. Methodology to evaluate the thermal insulation performance

To evaluate the thermal insulation performance of the alternative products processed with granulated corncob, the same experimental methodology was adopted. In this section, the facility, equipment, and calculation methodology are introduced and described.

4.1. Facility

The experimental work to determine the heat transmission coefficient (U) of the different case studies using the granulated corncob was carried out in a test room with the dimensions of 4.00×3.00×2.54 m (length×width×height). This test room has one of the façades oriented to the West and the other to the North. The existence of windows on the North façade allowed performing these tests due to the replacement of the window by the different samples analyzed.

The in situ thermal behavior assessment is valid if certain experimental conditions are achieved. These include the fact that the test samples have to be placed in an orientation that does not allow the solar radiation incidence and that allows the protection from the rain. It is also necessary to ensure the existence of a temperature gradient approximately constant between the inside and the outside of the test room to allow that the heat flow always occurs in the same way (from inside to outside during the winter and from the outside to inside the room during the summer). In the experimental procedure used in this research work, both conditions were guaranteed, contributing to the reliability of the results [14].

4.2. Equipment

In terms of equipment, a heat flux meter system was adopted as well as two thermohygrometric devices. A domestic heat device was also used to guarantee a constant temperature inside the test room. Each heat flux meter system is composed of two heat flux sensors (**Figure 4a**, detail I), four superficial temperature sensors (**Figure 4a**, detail II), a data logger (**Figure 4b**), and a computer [17]. The heat flux sensors allow to measure the heat flow through an element if there is a temperature gradient between the outdoor and indoor environments. Two superficial temperature sensors were used as a complement and as a reference to the heat flux sensor data and allowed to obtain the temperature values on the inner surface of the analyzed sample (**Figure 4a**). In **Figure 4c**, the adopted thermohygrometric device, which includes a temperature sensor (**Figure 4c**, detail I) and a relative humidity sensor (**Figure 4c**, detail II), is shown [14]. As mentioned above, two thermohygrometric devices were used, one to measure the temperature and humidity indoor and the other to measure the same parameters outdoor. In **Figure 4a**, a corncob particleboard sample is under this thermal behavior procedure.

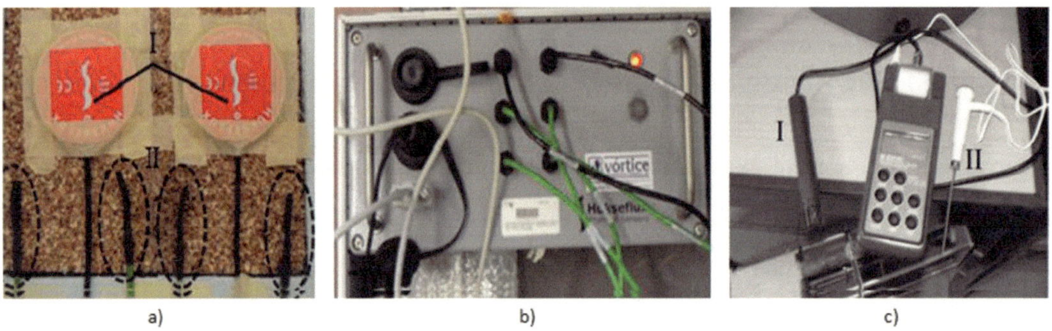

Figure 4. Equipment: (a) heat flux (I) and superficial temperature sensors (II), (b) data logger, and (c) thermohygrometric device: temperature (I) and relative humidity (II) probes.

4.3. Calculation methodology

According to ISO 9869 [17], the thermal transmission coefficient (U) of a material or a building system can be quantified given the heat flow that occurs through the element when submitted to a temperature differential, as presented in Expression (1):

$$U(ntotal) = \frac{\sum_{n=1}^{ntotal} q(n)}{\sum_{n=1}^{ntotal} \left(Ti(n) - Te(n)\right)} \quad (1)$$

where $q(n)$ is the heat flow across the element in the moment n; $Ti(n)$ and $Te(n)$ are the interior and the exterior temperatures in the moment n, respectively; $ntotal$ is the total number of moments in which the data were collected [14].

The use of the heat flux meters 1 and 2 allows obtaining the two heat flow values: $q1(n)$ and $q2(n)$. Considering the values of the temperature differential between the interior and the exterior environments, $Ti(n)$ and $Te(n)$, respectively, it is possible to estimate the thermal transmission coefficients, $U1(ntotal)$ and $U2(ntotal)$, respectively, by applying Expression (1). The thermal transmission coefficient of the element, $U'(ntotal)$, is the average value of $U1(ntotal)$ and $U2(ntotal)$, as shown in Expression (2) [14]:

$$U^{`}(ntotal) = \frac{U1(ntotal) + U2(ntotal)}{2} \tag{2}$$

According to ISO 9869 [17], a minimum of 3 days test duration is required if the temperature is stable around the heat flow-meters. Otherwise, the required duration may be 7 days or even more. This requirement intends to stabilize the heat flux that occurred through the element and it depends on the thermal inertia and the heat storage capacity of the building component. In the case of processed products based on granulated of corncob, the thermal inertia may be considered low and, consequently, a minimum of 3 days test duration may be acceptable.

There is a certain error related to the fact that the thermal transmission coefficient is obtained experimentally by the referred methodology. In this case, several aspects require to be taken into account to reduce this error, such as the calibration of the heat flux meters and the temperature sensors (there is ~5% error when these instruments are calibrated), the accuracy of the data acquisition system equipment, the difficulty of guarantying a perfect contact between the sensor and the surface of the sample (a 5% error may arrive), the operational error of the heat flux meter sensors due to the change of the isothermal curves, temperature, and heat flux changes, among others. Additionally, the error can be mitigated by increasing the number of sensors, stabilizing the interior temperature, and using a dynamic analysis. Thus, an error ranging from 14% to 28% is expected when the thermal coefficient transmission is estimated by the experimental procedure. However, in the present study cases, the aspects mentioned above were minimized. Therefore, the assessed $U'(ntotal)$ may be affected by an error of 14%. This error was considered in the analysis of the obtained results.

5. Granulated corncob preparation and its thermal insulation ability

To potentiate the application of corncob (**Figure 2a**) as a raw organic material in a product manufacturing process, it may be convenient to granulate it previously. Thus, it is important to apply an adequate granulating process that has to be able to maintain the important material properties of the corncob, in particular, its alveolar microstructure. For instance, an inadequate granulating process may damage the microstructure of the corncob by crushing or ripping the material. According to Pinto, Paiva and Faustino [18–20], a currently applied hammer mill in farms was the adopted device to granulate corncob (**Figure 5**).

Figure 5. Hammer mill: (a) general view, (b) lateral view, and (c) grid.

However, some difficulties emerged in what concern to obtain the specific granulometries of granulated corncob. Therefore, an alternative device was proposed, which consists of using a cutting mill device. In this case, the cutting mill device is provided with a range of sieves sized from 0.25 to 20 mm (**Figure 6**).

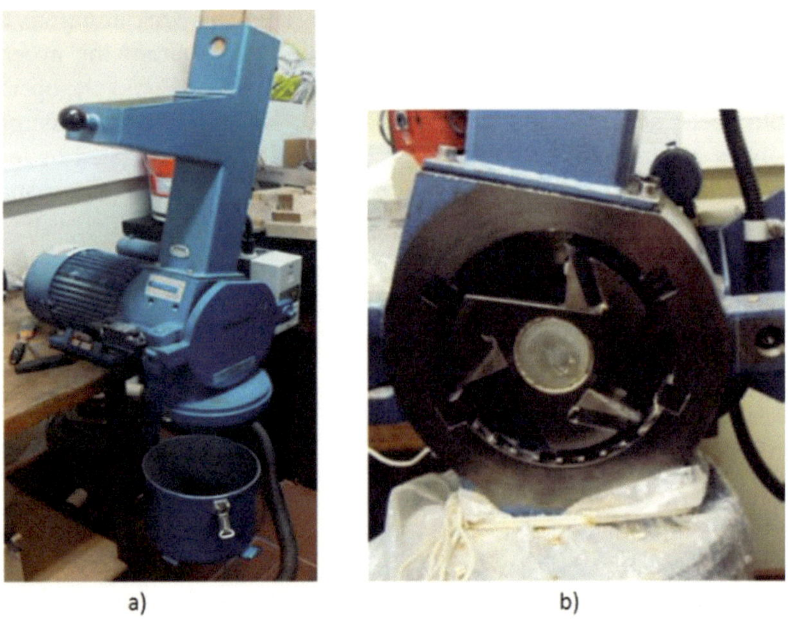

Figure 6. Cutting mill device: (a) general view and (b) cutting blade.

The granulated corncob obtained by this last process is shown in **Figure 7**.

Figure 7. Granulated corncob.

Granulated corn cob was then tested in terms of thermal insulation performance as shown in **Figure 8**, where the sensors used to measure the heat flux and the interior superficial temperatures can be observed.

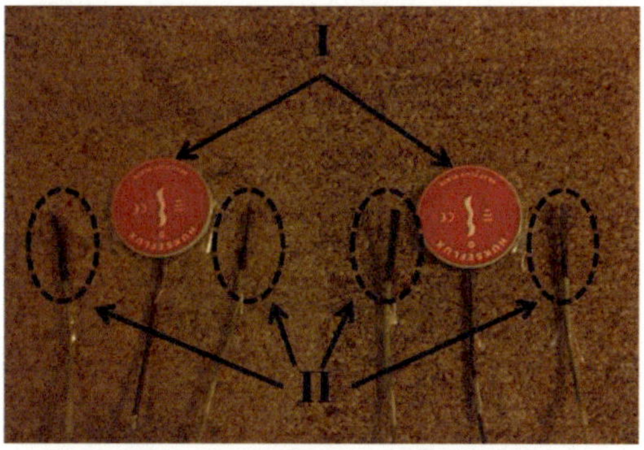

Figure 8. Granulated corncob sample under thermal insulation performance test (I: heat flux sensors and II: superficial temperature sensors).

The measured experimental data (temperatures and heat fluxes) during the test performance is presented in **Figure 9**. The temperature in the interior of the confined room was the temperature that was maintained approximately constant hanging around 25°C using a domestic heating device. The heat flow through the material sample [$q1(n)$ and $q2(n)$] was measured continuously (10-minute timing interval) using two heat flux sensors (**Figure 8**) for approximatly 7 consecutive days (n). The variation of the heat flow curves through the granulated corncob sample was in accordance with the one that characterizes the differential between the indoor and outdoor temperatures (**Figure 9**). Also, the temperature obtained outside was in accordance with the values that are usually obtained in the North of Portugal for this time of the year. These results allowed to conclude that the experimental test occurred as expected. The thermal transmission coefficient (*U*) was determined using Expression (1) and the values of the variables are presented in **Figure 9** [6].

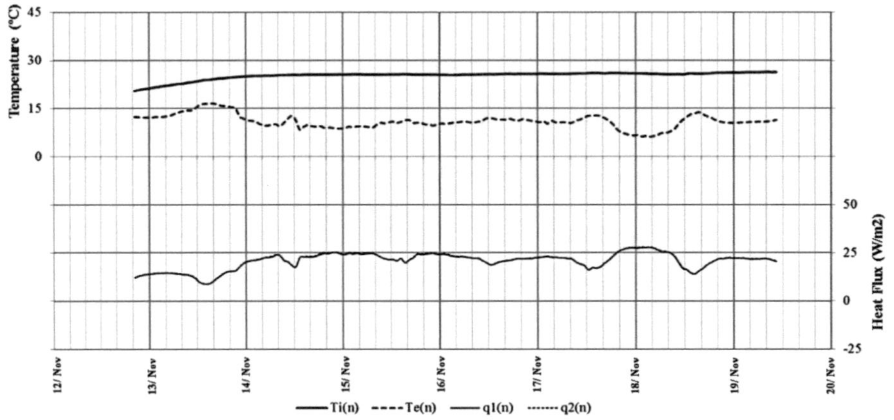

Figure 9. Granulated corncob. Indoor and outdoor temperatures [*Ti*(*n*) and *Te*(*n*)] and heat flow [*q*1(*n*) and *q*2(*n*)]. November 2014.

Figure 10 presents the thermal transmission coefficient (*U*) variation in the end of the measured period after its stabilization. The value of *U* obtained for the granulated corncob sample was 1.45 W/m^{2}°C. Furthermore, the thermal conductivity (λ) of the sample was also estimated based on these experimental results and the value obtained was 0.058 W/m°C. Thus, granulated corncob seems to offer proper thermal insulation because, according to ISO 9869 [17], the thermal conductivity of a conventional industrialized thermal insulation product must be less than 0.065 W/m°C.

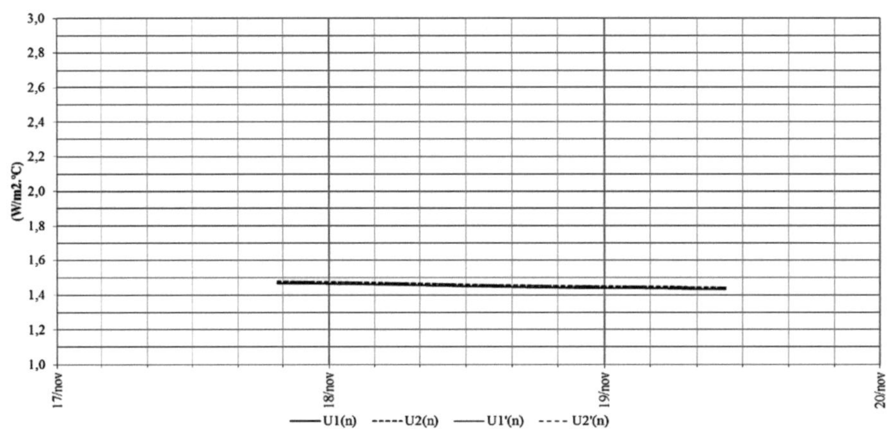

Figure 10. Granulated corncob. Thermal transmission coefficient (*U*) variation during the end of the measured period. November 2014.

6. Case studies of products based on corncob

As stated earlier, three alternative processed products based on corncob are presented here. In this respect, granulated corncob is first introduced followed by case studies I–III, which are related with corncob particleboard, lightweight concrete for nonstructural purposes, and lightweight concrete masonry unit CMU, respectively. The thermal insulation performance of these materials is evaluated by the experimental methodology described in the previous section.

6.1. Case study I – Granulated corncob particleboard

A corncob particleboard is the first processed product presented in this chapter. The particleboards were obtained through a simple manufacturing process that consists of binding the granulated corncob with wood glue [14]. This process included mixing up of the granulated corncob with wood glue, moulding, natural curing and unmoulding. A sample of a 3-cm-thick corncob particleboard (**Figure 11**) was tested in terms of thermal insulation according to the set-up described in Section 4.

Figure 11. A sample of a 3-cm-thick corncob particleboard.

In this case study, the thermal insulation test was performed for 6 consecutive days in March 2011. The measured experimental data (temperatures and heat fluxes) are presented in **Figure 12**. The temperature in the interior of the confined room was also kept nearly constant (23°C) using a domestic heating device. This temperature was always higher than the exterior temperature. These data allowed estimating the thermal transmission coefficient (U) by applying Expression (1). The value of U obtained for a 3-cm-thick granulated corncob particleboard was 2.14 W/m^2°C. A value of 0.101 W/m°C of the thermal conductivity was estimated. We believe that the thermal insulation performance of this product may be improved by refining its manufacturing process or increasing its thickness.

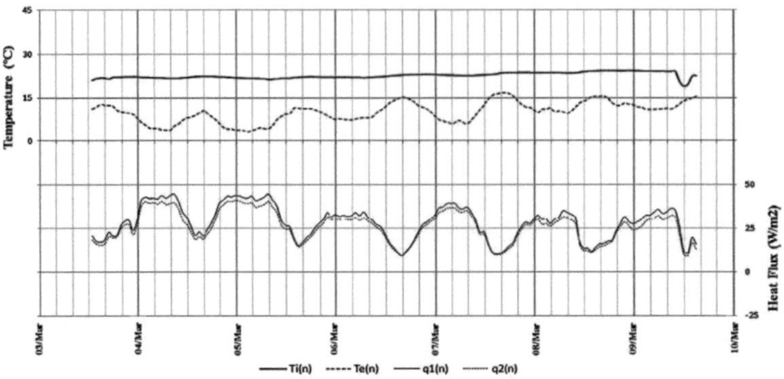

Figure 12. Case study I: corncob particleboard. Indoor and outdoor temperatures [*Ti*(*n*) and *Te*(*n*)] and heat flow [*q*1(*n*) and *q*2(*n*)]. March 2011.

Considering the values of the heat transmission coefficient (*U*) and thermal conductivity (*l*) estimated in the experimental work for the granulated corncob and (**Figure 10**) and for the corncob particleboard (**Figure 12**), it can be concluded that these values are similar to the ones that characterize the thermal insulation materials currently applied in the building industry. In **Table 1**, the thermal conductivity (*l*) is presented for different materials.

Insulation materials	Thermal conductivity (λ; W/m°C)
Granulated corncob	0.058
Corncob particleboard	0.101
EPS	0.04
XPS	0.032
Polyurethane	0.023
Cork (granules)	0.032–0.045
Glass wool	0.039
Rock wool	0.037
Expanded clay	0.103–0.108

Table 1. Thermal conductivity (λ) values of the corncob and current insulation materials.

6.2. Case study II – Lightweight concrete for nonstructural purposes based on granulated corncob

A lightweight concrete for nonstructural purposes based on granulated corncob was studied and proposed [21] (case study II). A weight ratio of 6:1:1 [i.e. lightweight aggregate (LWA) of granulated corncob/Portland cement/water] was applied, as it is the ratio used for the regularization layer of expanded clay concrete in the Portuguese construction. In **Figure 13**, the different stages of the corncob concrete sample processing are presented: adding the components (**Figure 13a**), curing process (**Figure 13b**), and the unmoulding step (**Figure 13c**). **Figure 13b** are also presents some corncob concrete samples used in the compression experimental test (**Figure 13b, I**) and corncob concrete samples used in the thermal performance experimental study (**Figure 13b, II**, and **Figure 13c**).

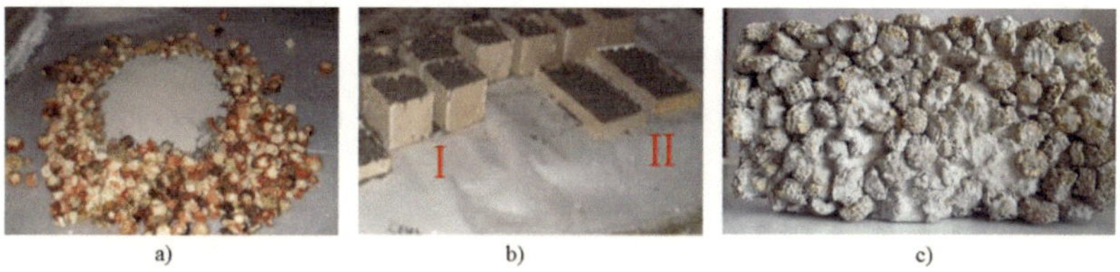

Figure 13. Manufacturing steps of lightweight concrete for nonstructural purposes based on corncob aggregate (case study III): (a) adding, (b) curing, and (c) unmoulding [21].

The experimental data measured during the thermal insulation behavior tests are displayed in **Figure 14**. It is worth to note that, in this case, lightweight concrete for nonstructural purposes based on expanded clay aggregate samples were also manufactured and tested to work as reference in terms of comparison. In this case, the test was performed for 5 days. The temperature of the test room was stabilized at around 20°C after 48 h (ΔT stabilizing; **Figure 14**). In what concerns to be the exterior temperature, it was verified that its values followed the expected oscillation during a day time (e.g. $\Delta Tnight$ and $\Delta Tday$; **Figure 14**). The analysis of the indoor and outdoor temperature values also showed that the outdoor temperature was always lower than the indoor temperature, which was also expected for that period of the year in that Portuguese region. Thus, those results lead to the conclusion that adequate thermal gradients were guaranteed (e.g. details I and II; **Figure 14**) to estimate the thermal transmission coefficient (*U*). Based on these experimental results and applying Expression (1), the *U* value of the lightweight concrete for nonstructural purposes based on corncob aggregate was possible to quantify and the respective value was 1.99 W/m^2°C [21].

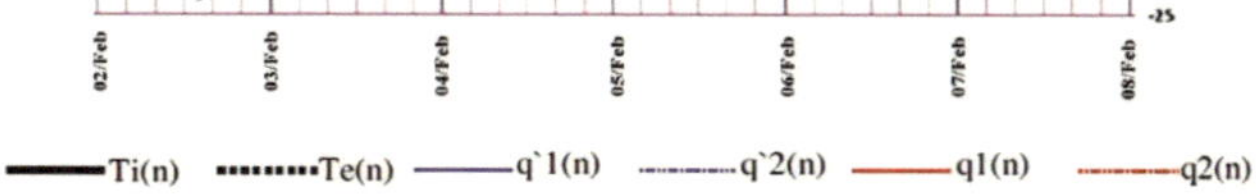

Figure 14. Case study II: lightweight concrete for nonstructural purposes. Interior [*Ti*(*n*)] and exterior [*Te*(*n*)] temperatures. Heat flow across the corncob [*q*′1(*n*) and *q*′2(*n*)] and the expanded clay [*q*1(*n*) and *q*2(*n*)] concrete samples. February 2011.

6.3. Case study III – Lightweight concrete masonry unit based on processed granulated corncob

A research work was performed to assess the potential application of PCC as an alternative LWA for the manufacturing process of lightweight CMU [22]. Therefore, CMU-PCC was prepared in a factory using a typical lightweight concrete mixture for nonstructural purposes (**Figure 15a**). Medium sand (MS; 0.0–4.0 mm), coarse sand (LS; 0.8–3.0 mm), gravel (G; 2.0–6.0 mm), Portland cement 32.5 N (C), LWA, and water (W) were the constituents considered in this research for manufacturing lightweight CMU. The respective adopted mixture to manufacture a CMU is presented in **Table 2**.

a) b)

Figure 15. Lightweight CMU: (a) granulated corncob (PCC) and (b) expanded clay (EC).

	LS	G	C	LWA	W
1.530	1.836	3.060	1.326	1.326	1.326

Table 2. Adopted mixture in the manufacturing process of CMU (kg) [23].

In this case, the particles of corncob were previously covered with cement paste. This procedure intended to solve some material limitations, such as high level of water absorption of the granulated corncob, slow drying process, and low compressive strength of the lightweight concrete produced [21].

As adopted in case study II, samples of CMU based on expanded clay (CMU-EC) were also prepared and tested to work as reference in terms of comparison (**Figure 15b**).

The CMU were then tested in terms of thermal insulation behavior and according to the experimental procedure described in Section 4. **Figure 16** shows the CMU under this test. Thus, CMU-PCC and CMU-EC were tested at the same time and under the same thermohygrometric conditions.

http://dx.doi.org/10.5772/62339

Figure 16. Thermal insulation test of CMU (exterior view). March 2014.

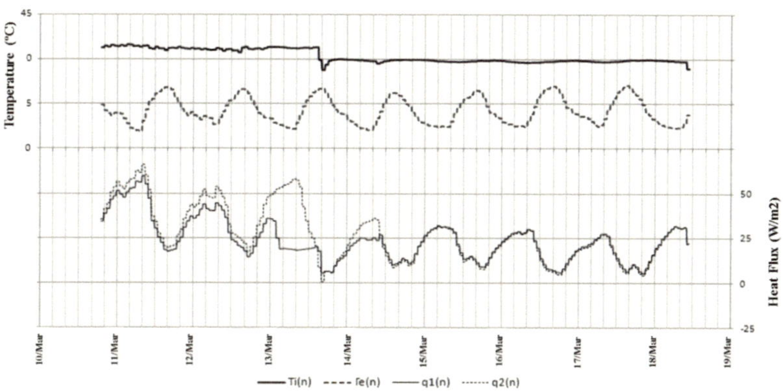

Figure 17. Case study III: CMU-PCC. Interior [*Ti*(*n*)] and exterior [*Te*(*n*)] temperatures. Heat flow across the sample [*q*1(*n*) and *q*2(*n*)]. March 2014.

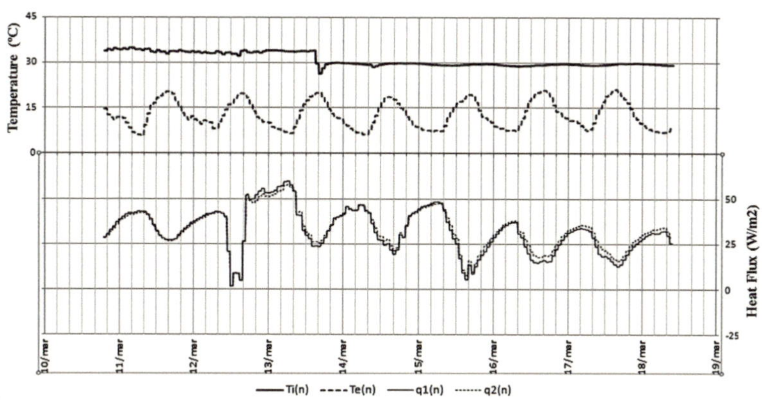

Figure 18. Case study III: CMU-EC. Interior [*Ti*(*n*)] and exterior [*Te*(*n*)] temperatures. Heat flow across the sample [*q*1(*n*) and *q*2(*n*)] March 2014.

The experimental data measured for CMU-PCC is presented in **Figure 17**. The test duration was 8 days.

The quantified thermal transmission coefficient (*U*) of CMU-PCC was 1.15 W/m^2°C.

Figure 18 presents the heat flux variation obtained during the test period for the lightweight CMU-EC. These values lead to a thermal transmission coefficient of 1.75 W/m^2°C, which is high than the value obtained for the CMU-PCC. The results show that the introduction of the granulated corncob in the concrete unit increases its thermal insulation performance.

7. Conclusions

An example of a Portuguese traditional application of corncob as a building material was presented.

Some material properties, such as the macrostructure, microstructure, density, water absorption, and fire resistance of corncob, were indicated and compared to currently applied thermal insulation materials (e.g., XPS, EPS, cork, and expanded clay). It was highlighted that the corncob material may be very heterogeneous, discontinuous, and anisotropic. It was also stated that there must be interesting similarities between corncob and cork.

Some considerations about the advantages of granulating corncob were done and some information about possible granulating processes was delivered.

An alternative expedite experimental set-up based on ISO 9869 [17] was proposed to evaluate the thermal insulation performance of the different applications of the granulated corncob. A test room with a constant interior temperature and a façade oriented north to avoid the solar radiation intensity and the effect of the rain in the testing sample was necessary. To estimate the thermal transmission coefficient of the different samples, a heat transfer system was used. It was composed of two heat flux sensors to measure the heat flow across the sample and four superficial temperature sensors to obtain the interior superficial temperature of the sample. Two thermohygrometric devices were used to obtain the temperature and the relative humidity values of the indoor and outdoor environments. The research work developed led to the conclusion that this experimental procedure is accurate and allowed to perform thermal performance analysis under real climatic conditions, to use real samples, to test several samples simultaneously, and to monitor the thermal behavior of a construction element continuously for several days [14].

Experimental data concerning the thermal insulation of granulated corncob were given. In addition, the value of the thermal transmission coefficient of this material was estimated as 1.45 W/m^2°C. The respective thermal conductivity value was 0.058 W/m°C.

On the other hand, a corncob product of 3-cm-thick corncob particleboard (case study I) was introduced and its thermal insulation behavior was also studied. A 2.14 W/m^2°C thermal transmission coefficient and a 0.101 W/m°C thermal conductivity for this product based on granulated corncob were quantified.

In addition, granulated corncob is also proposed as an alternative organic aggregate of lightweight concrete and as an alternative to expanded clay, cork, and EPS, among other possibilities. The proposed corncob lightweight concrete (case study II) was processed using the ratio of 6:1:1 (granulated corncob/Portland cement/water) and the results allowed to conclude that the density and the thermal properties of this alternative lightweight concrete are in accordance with the current expanded clay concrete properties [21].

Finally, lightweight CMU-PCC was presented as other product based on the agricultural waste material under this research (case study III). In this respect, the experimental results obtained so far are very promising, in particular, in terms of its thermal insulation behavior. In fact, experimental data concerning this ability are delivered in this chapter and also the thermal transmission coefficient (*U*) of CMU-PCC was estimated as being equal to 1.15 $W/m^{2}{}^{\circ}C$.

The information presented is a result of several works that were intended to find possible applications of corncob as a raw building material and having a thermal insulation perspective. These research works allowed to conclude that this organic and agricultural waste has the potential to be used as a component of thermal insulation materials, systems, or products.

Acknowledgements

This work was partially supported by the FCT (Portuguese Foundation for Science and Technology) through the project PEst-OE/ECI/UI4082/2013 (C-MADE).

We also acknowledge the Forest Products Laboratory of the Forest Department, the Civil Engineering Laboratories of the Engineering Department, and the Electronic Microscopy Unit of the Chemistry Department of the University of Trás-os-Montes Alto Douro.

A special thanks to Daniel, Helder, Elisabete, Nuno, and Juliana.

Author details

Jorge Pinto[1,2*], Ana Briga Sá[1,2], Sandra Pereira[1,2], Isabel Bentes[1,2] and Anabela Paiva[1,2]

*Address all correspondence to: tiago@utad.pt

1 ECT-School of Science Technology, University of Trás-os-Montes e Alto Douro, UTAD, Quinta de Prados, Portugal

2 C-MADE-Centre of Materials and Building Technologies, University of Beira Interior, Portugal

References

[1] Younquist J, English B, Spelter H, Chow P. Agricultural fibers in composition panels. In: Proceedings of the 27th International Particleboard/Composite Materials Symposium, 30–31 March, Pullman, Washington, 1993, pp. 133–152.

[2] Chow P. Dry Formed Composite Board from Selected Agricultural Residues. World Consultation on Wood Based Panels, Food and Agriculture Organization of the United Nations, New Delhi, India, 1974.

[3] Lertsutthiwong P, Khunthon S, Siralertmukul K, Noomun K, Chandrkrachang S. New insulating particleboards prepared from mixture of solid wastes from tissue paper manufacturing and corn peel. Bioresource Technol 99 (2008) 4841–4845.

[4] Khedari J, Nankongnab N, Fotios S. Agricultural waste materials as thermal insulation for dwellings in Thailand: preliminary results. In: PLEA 2008: 25th Conference on Passive and Low Energy Architecture, 22–24 October, Dublin, Ireland, 2008.

[5] Stone N. Thermal performance of straw bale wall systems. Ecol Build Netw (EBNet) 2003:1–7.

[6] Pinto J, Cruz D, Paiva A, Pereira S, Tavares P, Fernandes L, Varum H. Characterization of corn cob as a possible raw building material. Constr Build Mater 2012;34:28–33.

[7] ANPROMIS-Associação Nacional dos Produtores de Milho e Sorgo. Available at: <http:// www.anpromis.pt/>.

[8] Pinto J, Paiva A, Varum H, Costa A, Cruz D, Pereira S, et al. Corn's cob as a potential ecological thermal insulation material. Energy Build 2011;43:1985–90.

[9] Guan J, Hanna M. Functional properties of extruded foam composites of starch acetate and corn cob fiber. Ind Crop Prod 2004;19:255–69.

[10] Panthapulakkal S, Sain M. Agro-residue reinforced high-density polyethylene composites: fiber characterization and analysis of composite properties. Composites Part A 2007;38:1445–54.

[11] Dowling A, Mathia J. Experimental determination of the insulating ability of corn by-products. J Sustain Agr 2007;30:15–27.

[12] Adesanya D, Raheen A. Development of corn cob ash blended cement. Constr Build Mater 2009;23:347–52.

[13] Pinto J, Paiva A, Costa A, Tavares P, Fernandes L, Murta A, Varum H. Building with sugar and corn. In: Advances in Production Management Systems-APMS2010. Competitive and Sustainable Manufacturing Products and Services. Cernobbio, Como, Italy, 11–13 October 2010.

[14] Paiva A, Pereira S, Sá A, Cruz D, Varum H, Pinto J. A contribution to the thermal insulation performance characterization of corn cob particleboards. Energy Build 2011. doi: 10.1016/j.enbuild.2011.11.019.

[15] Pinto J, Paiva A, Varum H, Costa A, Cruz D, Pereira S, Fernandes L, Tavares P, Agarwal J. Corn's cob as a potential ecological thermal insulation material. Energy Build 2011;43:1985–90.

[16] Santos C, Matias L. Coeficientes de Transmissão Térmica de Elementos da Envolvente dos Edifícios-ITE 50, Coleção Edifícios-ITE 50. LNEC, Lisbon, Portugal, 2006. ISBN: 978-972-49-2065-8 (in Portuguese).

[17] ISO 9869. Thermal Insulation: Building Elements—In Situ Measurement of Thermal Resistance and Thermal Transmittance. International Organization for Standardization (ISO), 1994.

[18] Pinto J, Cruz D, Paiva A, Pereira S, Tavares P, Fernandes L, Varum H. Characterization of corn cob as a possible raw building material. Constr Build Mater 2012;34:28–33. doi: 10.1016/j.conbuildmat.2012.02.014.

[19] Paiva A, Pereira S, Sá A, Cruz D, Varum H, Pinto J. A contribution to the thermal insulation performance characterization of corn cob particleboards. Energy Build 2012;45:274–9. doi: 10.1016/j.enbuild.2011.11.019.

[20] Faustino J, Pereira L, Soares S, Cruz D, Paiva A, Varum H, Ferreira J, Pinto J. Impact sound insulation technique using corn cob particleboard. Constr Build Mater 2012;37:153–9.

[21] Pinto J, Vieira J, Pereira H, Jacinto C, Vilela P, Paiva A, Pereira S, Cunha V, Varum H (2012). Corn cob lightweight concrete for non-structural applications. Constr Build Mater 2012;34:346–51. doi: 10.1016/j.conbuildmat.2012.02.043.

[22] Faustino J, Silva E, Pinto J, Soares E, Cunha V, Soares S. Lightweight concrete masonry units based on processed granulate of corn cob as aggregate. Mater Constr 2015;65:e055. doi: 10.3989/mc.2015.045.

Chapter 3

Thermal Insulation Material Based on "Jute"

Sanjoy Debnath

Additional information is available at the end of the chapter

http://dx.doi.org/10.5772/63223

Abstract

Among the different natural fibres, jute is a less expensive fibre, annually renewable, and commercially available compared to other natural fibre crops. This jute is mostly cultivated in India and Bangladesh. More than a century, this fibre is well known as packaging (sacks), hessian, and carpet backing. Since 1950s, the synthetic fibres slowly took the market share of conventional jute textiles due to their low cost and high production speed. As far as suitability of the insulating material is concerned, it has high potential of using as three types of insulation (thermal, sound, and electrical). This present chapter gives emphasis on the basic methods of measuring jute based thermal insulation materials in different application areas. Apart from its evaluation methods, special attention has been made on the important factors affecting the thermal insulation behaviours of the jute-based textile materials. Focusing the needs of the industry, present chapter also covers the future aspects regarding the insulation application from jute-based materials.

Keywords: technical textiles, insulation materials, jute textile material, thermal insulation, woven and nonwoven structures

1. Introduction

Among the different fibre crops, jute is one of the oldest cultivated fibre crops in India. Jute is mostly cultivated in the eastern part of India, and stands highest production in the world, used popularly as technical textiles over the centuries. Jute fibre is used for reinforcement of rural mud house. Jute sacks are used as thermal insulating material [1], and for domestic animals like cattle, goat, pet dog, etc. Apart from these, it is the cheapest fibre crop available commercially in bulk quantities as of today. As far as the properties of jute fibre are concerned, it has both good characters as well as unwanted properties. Basically, this fibre is a mesh like structure which provides good coverage, good tensile strength, provides toughness and durability, less elongation at break, ensures dimensional stability, and natural colour which is ethnic in nature. Unlike any other fibres, the drawbacks of jute fibre crop are high surface roughness and prickliness, low extension at break, and coarseness, which restricts its use in textile garment.

Apart from these properties, jute-based materials have also properties like thermal, sound, and electrical insulation materials, out of which application in thermal insulation area is more popular [1, 2]. As per the usage of the material, insulation material can be classified as wearable textile and non-wearable textiles. Wearable textiles are those which are worn by any person either in direct contact with the skin or used as secondary clothing like jacket, protective clothing [3], gloves, etc. On the contrary, non-wearable materials are those that are not used directly by human beings, rather they are used in an indirect way like, insulation carpet, floor mat, insulation used in covering the electrical cable as protection material, roof top covering, wall coverings, etc. Now-a-days, jute-based materials are being used in the form of fibres, yarn, fabric, and composite media. There are researches where the method of measurement of insulation property, and the effect of such properties on different external parameters are demonstrated.

Keeping these in view, warm clothes have been designed and developed using jute-based fibres and yarns. Thermal insulation is one of the essential properties for any warm fabrics [3–6]. Judicious modifications on the fibre/yarn structure are one of the important parts as far as the thermal insulating material is concerned. The thermal insulation related properties mainly depend on the availability of amount of air pores in the textile structure. The static air trapped in fabric pores, makes the fabrics act as thermal insulating media [2]. As per the sound insulation is concerned, it mostly depends on the material surface morphology. Here, the morphology indicates the surface roughness, voids on the surface of the material, compactness of the material, intensity of the roughness, material structure (woven/nonwoven), etc.

Out of these three (thermal, sound, and electrical) basic types of jute-based insulation materials, major contribution has been documented in the area of thermal insulation. Hence, the major emphasis in this chapter has been imparted on the characterisation of the thermal insulation of the jute-based material factors affecting the thermal insulation of those materials and possible applications of jute based thermal insulation materials.

2. Evaluation of thermal insulation

The thermal resistance of textile material is generally defined as the ratio of the temperature difference between the two surfaces of the textile fabric material to the rate of flow of heat per unit area normal to the surfaces. This is analogous to the electrical resistance in case of current flow through electrical conductor. In Disc method, an application of Lee's disc apparatus to textiles has been used to evaluate thermal resistance of needle-punched nonwoven fabric samples. The material under test is kept between two metal discs surfaces from which, one has known thermal resistance. In a steady condition, the temperature drop across the metal disc with known value of thermal resistance and across the material under test is measured, and

http://dx.doi.org/10.5772/63223

from the values obtained, the thermal resistance of the specimen is determined by the following techniques [4].

Let TR_k and TR_s be the thermal resistance of the known disc and the sample under test respectively. Let t_1 be the temperature registered by the lower surface of known disc, t_2 be the temperature registered by the lower surface of the sample under, and t_3 be the upper surface of the sample under test. Assuming constant rate of flow of heat at steady state condition, the TR_s is computed from the following formula in degrees Kelvin square metre per Watt:

$$\frac{t_1 - t_2}{TR_k} = \frac{t_2 - t_3}{TR_s}, \text{ or } TR_s = TR_k \times \frac{t_2 - t_3}{t_1 - t_2} \quad (1)$$

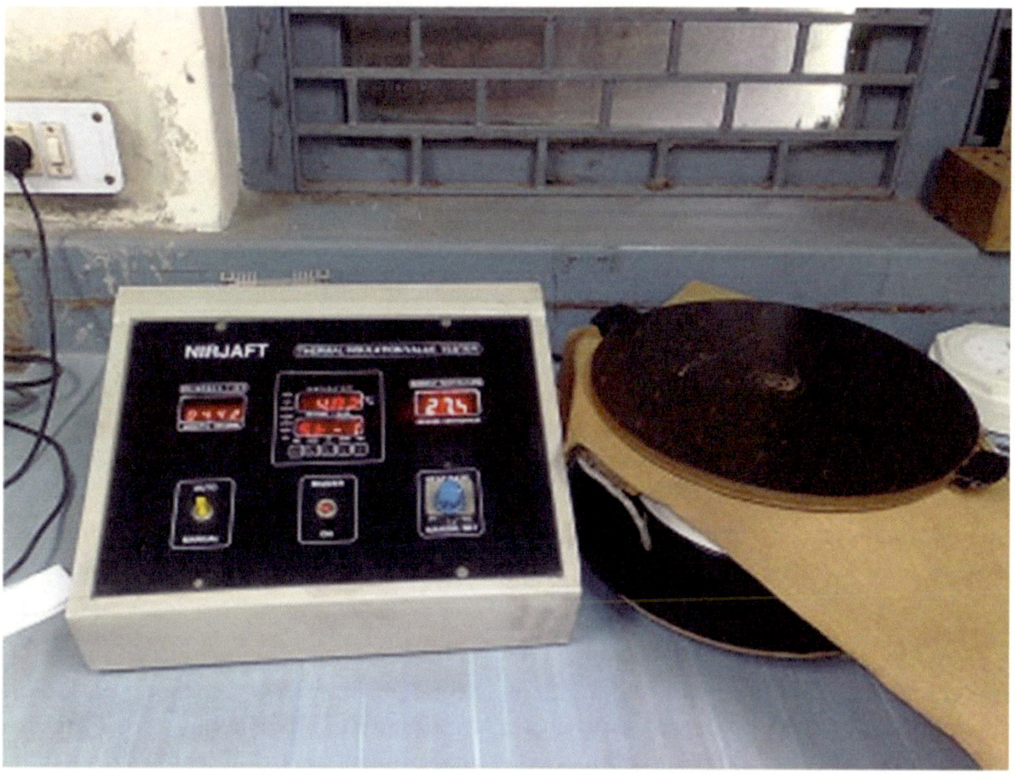

Figure 1. Instrument for measuring the thermal resistance of fabrics.

In this experiment, guarded two-plate thermal resistance instrument has been used to measure the thermal resistance of jute-based needle-punched fabrics (**Figure 1**) [4–6]. The thermal resistant instrument is based on a microprocessor and provides automatic results of thermal resistance value in 'tog'. The area of the test specimen used is 706.85 cm^2 (diameter 30 cm). The test is non-destructive and the process of sample preparation is free from human error. Thermal insulation of each fabric sample is measured randomly at five different places under a pressure of 0.3352 kPa. Average of five readings was considered and the coefficient of variation of the readings was <2%. All the fabric materials must be conditioned in the standard atmospheric condition prior to evaluation of the thermal insulation property [7].

Specific thermal resistance (STR_s) value is used to compare the thermal resistance of different nonwoven fabric samples. STR_s values of all the samples are determined using the following equation [4]:

$$STR_s = \frac{TR_s}{T_0} \tag{2}$$

where STR_s is the specific thermal resistance in K m^2/W; TR_s, the thermal resistance value of fabric in K m^2/W; and T_0, the mean thickness in meter at 1.55 kPa pressure of the fabric sample.

3. Jute-based insulation materials and important factors affecting their insulation properties

Jute fibre has its inherent property of good thermal insulation. Different construction of jute-based textile materials further enhanced the performance and properties of insulation [8]. There are different applications where the jute based structures are used as insulating material like, warm garments, floor mat, carpet, soil temperature control in agricultural application, false ceiling, temporary partition wall, sound absorbing material in auditorium, etc. Depending upon the insulation requirement, different textile structures are being used like fibre, yarn, and fabric. Sometimes, composite structures are also used as particle board, and fibre reinforced board. Again in fabric, woven, nonwoven, and knitted structures are being used as insulation material. The following studies elaborate details on different possible insulation materials from jute-based textiles.

3.1. Thermal insulation behaviour of jute-based nonwoven fabrics

Different types of parallel laid and random laid needle punched and adhesive bonded nonwoven fabrics were prepared using blending of different fibre materials (polypropylene, acrylic, jute, woollenised jute, jute caddis, cotton, wool, ramie, pineapple leaf fibres, etc.). Two types of blending methods were used such as sandwich and homogeneous. Sandwich blending of polypropylene or acrylic with woollenised jute shows better thermal insulation compared to homogeneous blended materials as found by Debnath. They also found that nonwoven prepared out of woollenised jute-wool (2:1), woollenised jute-acrylic (2:1), and woollenised jute-pineapple leaf fibre (2:1) have better thermal insulation property. Air permeability and thermal conductivity of jute needle-punched nonwoven fabrics have been studied by Debnath et al. [3] and found that jute needle punched nonwoven has poor heat transmission. Furthermore, Box and Behnken factorial design was used to design and develop needle-punched nonwoven fabrics made from jute and polypropylene blends to study the effect of fabric weight, needling density, and blend proportion on thickness, thermal resistance, STR_s, air permeability, and sectional air permeability. Polypropylene fibre of 0.44 tex fineness, 80 mm length, and jute fibres of Tossa-4 grade were used to develop the jute-polypropylene blended needle-punched nonwoven. Some of the important properties of these jute and polypropylene fibres are presented in the **Table 1**.

Property	**Jute**	**Polypropylene**
Fibre fineness, tex	2.08	0.44
Density, g/cm^3	1.45	0.92
Moisture regain at 65% RH, %	12.5	0.05
Tensile strength, cN/tex	30.1	34.5
Breaking elongation, %	1.55	54.13

Table 1. Properties of jute and polypropylene fibres [4].

3.2. Preparation of jute-polypropylene blended thermal insulation nonwoven fabrics

Initially, the jute reeds were opened through a roller and clearer card. This produces almost mesh-free opened stapled fibre. The woollenised jute and polypropylene fibres are then hand opened separately and blended at three different blend proportions as per the **Table 2**. Considering the fibre droppings in different stages of woollenised jute fibres taken is 2% higher than presented in **Table 2** to maintain the target blend proportion. Then the blended materials were opened thoroughly by passing through one carding passage.

The blended fibres were then fed to the lattice of the roller and clearer card at a uniform and predetermined rate so that a web of 50 g/m^2 can be achieved. The fibrous web coming out from the card was fed to feed lattice of cross-lapper, and cross-laid webs were produced with cross-lapping angle of 20°. The web was then fed to the needling zone. The required needling density was obtained by adjusting the throughput speed.

As per the fabric weight (g/m^2) requirement, certain number of webs was taken and passed through the needling zone of the machine for a number of times, depending upon the punch density required. A punch density of 50 punches/cm^2 was applied on each passage of the webs, reversing the face of the web alternatively [4]. The fabric samples were produced as per the coded and actual levels of three variables (**Table 2**).

The depth of needle penetration was kept constant at 11 mm. For all webs, 15 × 18 × 36 × R/SP, 3½ × ¼ × 9 needles were used.

Fabric code	Levels of variables					
	X_1 level		X_2 level		X_3 level	
	Coded	Actual	Coded	Actual	Coded	Actual
1	−1	250	−1	150	0	60:40
2	−1	250	1	350	0	60:40
3	1	450	−1	150	0	60:40
4	1	450	1	350	0	60:40
5	−1	250	0	250	−1	40:60
6	−1	250	0	250	1	80:20
7	1	450	0	250	−1	40:60
8	1	450	0	250	1	80:20
9	0	350	−1	150	−1	40:60
10	0	350	−1	150	1	80:20
11	0	350	1	350	−1	40:60
12	0	350	1	350	1	80:20
13	0	350	0	250	0	60:40
14	0	350	0	250	0	60:40
15	0	350	0	250	0	60:40

X_1 – Fabric weight, g/m²; X_2 – Needling density, punches/cm²; and X_3 – Blend ratio (polypropylene: woollenised jute).

Table 2. Actual and coded values for three independent variables and the experimental design [4].

3.3. Effect of fabric weight, needling density, and blend proportion of jute-polypropylene blended needle-punched nonwoven on thermal resistance

It is found that the thermal resistance increases with the increase in fabric weight [4] significantly (p <s 0.05000 and positive correlation, $r = 0.82$) as obtained from **Table 3**. There is more prominent increase of thermal resistance value of the fabric with the increase in fabric weight at 150 needling density than is obtained at 350 punches/cm². With the increase in needling density within the experimental range, the thermal resistance has does not have any significant effect even with varying jute component in the blend from 40% to 60%. Optimum thermal resistance value of 8.5×10^{-2} K m²/W found at fabric weight of 430 g/m², needling density of 150 punches/cm², and jute content of 40% in the blend. The number of fibres per unit volume of the fabric increases with the increase in fabric weight, which results in higher fabric thickness and larger amount of voids in the fabric structure obtained. These ultimately increase the thermal resistance of the fabric while increase in fabric weight. On the contrary, while needling density increases, thermal resistance decreases significantly ($p < 0.05000$ and negative correlation, $r = -0.67$) as depicted from correlation matrix (**Table 3**). This is due to higher degree of consolidation occurred, and hence reduces voids in the structure. Since the air act as thermal insulating material, the fall in air pocket in the fabric structure reduces the thermal resistance of the jute blended fabric.

Variables	*FW*	N_p	*J%*	T	TR_s	STR_s	*AP*	*SAP*
FW	1.00	–	–0.00	0.50	0.51	0.28	–0.93	–0.75
N_p	0.00	1.00	0.00	–0.49	–0.67	–0.61	–0.11	–0.33
J%	–0.00	0.00	1.00	–0.39	–0.26	–0.02	–0.19	–0.43
T	0.05	–0.49	–0.39	1.00	0.82	0.29	–0.36	0.08
TR_s	0.51	–0.67	–0.26	0.82	1.00	0.78	–0.37	–0.02
STR_s	0.28	–0.61	–0.02	0.29	0.78	1.00	–0.22	–0.11
AP	–0.93	–0.11	–0.19	–0.36	–0.37	–0.22	1.00	0.89
SAP	–0.75	–0.33	–0.43	0.08	–0.02	–0.11	0.89	1.00

FW – Fabric weight, g/m²; N_p – Needling density, punches/cm²; *J%* – Jute proportion, T_0 – Fabric thickness, cm; TR_s – Thermal resistance × 10^{-2}, K m²/W; STR_s – Specific thermal resistance, K m/W; *AP* – Air permeability, cm³/cm²/s; SAP – Sectional air permeability, cm³/s/cm.
*Correlations are significant at $p < 0.05000$.

Table 3. Correlation matrix of variables [4].

Thermal resistance = $4.0520833 - 0.0114167X_1 - 0.0007917X_2 + 0.0558333X_3\ 0.0000079X_1^2 - 0.0000104X_2^2 - 0.0021979X_3^2 + 0.0000250X_1X_2 - 0.0002125X_1X_3 - 0.0001X_2X_3$ ($R = 0.9002$; $F_{9,5} = 15.04$)

3.4. Effect of fabric weight, needling density, and blend proportion of jute-polypropylene blended needle-punched nonwoven on specific thermal resistance

A study on specific thermal insulation behaviour of jute-polypropylene blended needle-punched nonwoven [4] show that the STR_s depends prominently on varying jute content levels of 20%, 40%, and 60% respectively (**Figure 2**). This study also reveals that with the increase in needling density, the STR_s decreases. They found that between needling density and STR_s, there is a significant ($p < 0.05000$) negative correlation ($r = -0.61$), shown in the correlation matrix (**Table 3**). Formation of consolidated structure occurs with the increase in needling density, as a result, the available air pockets reduces in the fabric structure. Again, with the increase in fabric weight, the number of fibres in unit area of the fabric increases, which increases the voids in the fabric structure. These ultimately influence the STR_s of the needle-punched nonwoven fabric. It is depicted from **Figure 2a**, that initially thermal resistance increases up to 375 g/m² of the fabric weight and later it decreases with further increase in fabric weight. Same trend has also been observed at 60% jute content level, but the decrease in trend of STR_s occurs at lower fabric weight (325 g/m²) as obtained from **Figure 2b**. This is due to reason that compared to polypropylene fibre, jute can easily form consolidated structure due to its poor resilience. Because of this, during higher level of needling and jute content levels, the fabric consolidation initially improves, and beyond certain fabric weight (325 g/m²) the bulkiness increases. Larger number of fibres available for each needle during needling with the increase in fabric weight means more number of fibres will be available to the needle barb during needling. Further increase from 325 g/m^2 of fabric weight, the incremental amount of fibres at needle barb is deficient to form better entanglement, which produce poor consolidation. Thus, with the increase in jute content (60%), the fabric consolidation occurs at 325 g/m^2 fabric weight (lower level) compared to that occurred at 40% jute content level (**Figure 2c**). Optimum STR_s value of 20.6 K m/W was obtained at 150 punches/cm^2 needling density and 400-450 g/m^2 fabric weight at lower jute content (40%) of jute-polypropylene blended needle-punched nonwoven (**Figure 2b**).

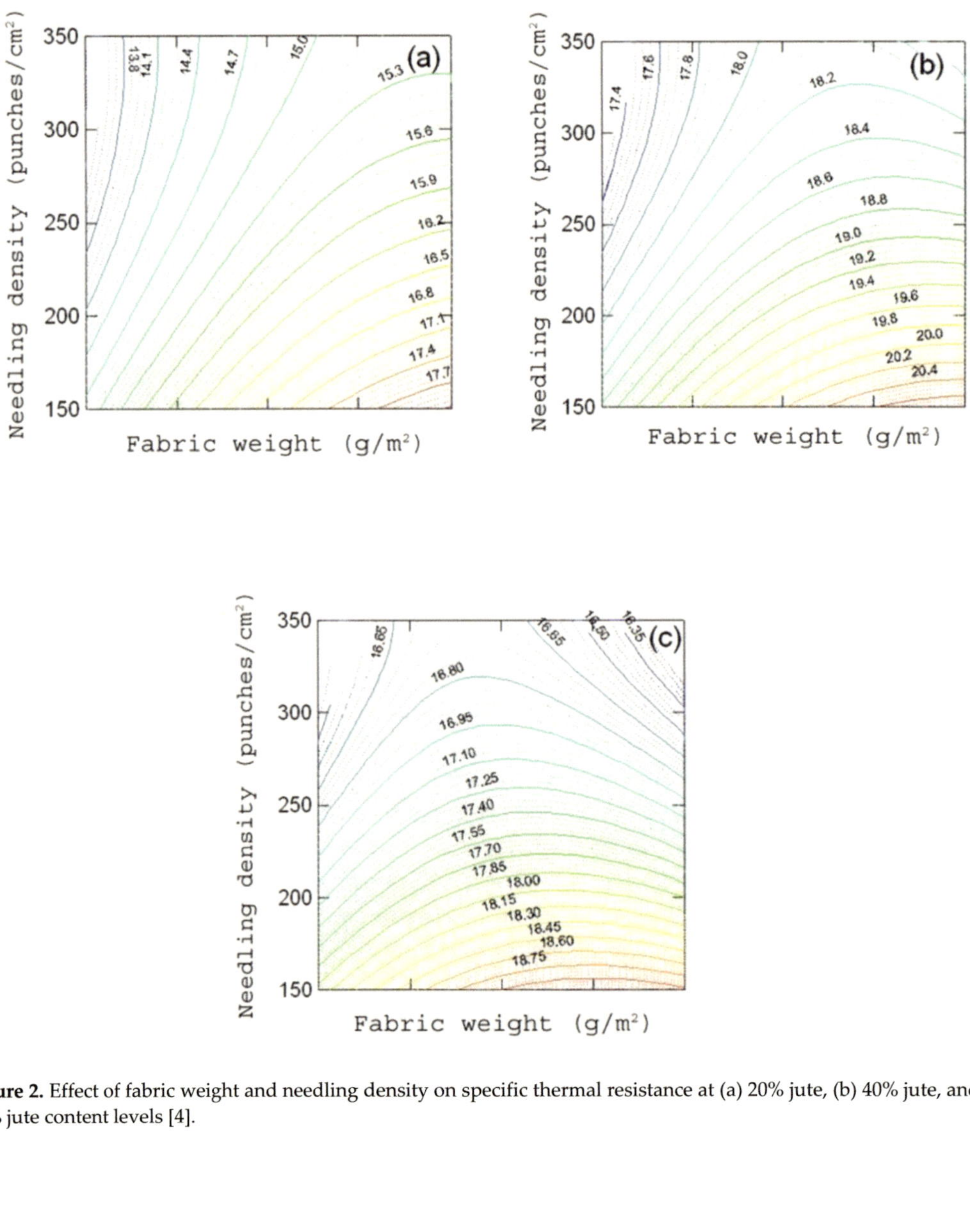

Figure 2. Effect of fabric weight and needling density on specific thermal resistance at (a) 20% jute, (b) 40% jute, and (c) 60% jute content levels [4].

http://dx.doi.org/10.5772/63223

Specific thermal resistance= $-2.3122917 + 0.0612292X_1 - 0.0160917X_2 + 0.5955833X_3 - 0.0000490X_1^2 + 0.0000452X_2^2 - 0.0056073X_3^2 - 0.0000365X_1X_2 \cdot 0.0002725X_1X_3 - 0.0002163X_2X_3$ ($R = 0.9327$; $F_{9,5} = 7.69$)

Furthermore, Yachmenev et al. [9], have discussed the thermal insulation properties of biodegradable, cellulosic-based nonwoven composites for automotive application. This work is aimed to the development of bio-composite from jute-based material aiming for automobile application. They developed moldable, cellulosic-based nonwoven composites with excellent thermal insulation properties, which were fabricated from kenaf, jute, flax, and waste cotton using recycled polyester and substandard polypropylene. The composites of these fibers have excellent shape, stability, and high tensile and flexural properties coupled with economic and environmental benefits. Four different types of designs with varying different cellulosic fibers, manufacturing techniques, and various ratios of vegetable–synthetic fibers were manufactured on laboratory-scale equipment. A Steady-State Heat Flow meter was used for the measurement of thermal conductivity and thermal transmittance of the samples of composites. The research findings show that thermal insulation properties of the cellulosic-based nonwoven composites vary significantly, depending on the type of the cellulosic fibers, the ratio of cellulosic fibers to synthetic fibers, and the resulting density of the composite [9].

3.5. Measurement of thermal insulation value and comparative study of different jute based materials

A simple method can be used to measure the thermal insulation value (TIV) of different textile materials based on jute and cotton fibres [8, 10–14]. Methods that are commonly used for measurement of TIV are the disc method, the constant temperature method, and cooling method. Out of these three methods, cooling method is the simplest compared to other two methods. In this method of measurement of thermal insulation, a hot body is wrapped with the fabric, and its rate of cooling is measured. The outer side of the fabric is exposed to air. In this experiment, the time taken by a hot body covered with the fabric sample (t_c) and without the sample (t_u) to cool through a particular temperature range under identical atmospheric conditions. To measure the thermal insulation with this method, a brass cylinder (45 cm length, 5 cm external diameter, and 2 mm thickness) closed at one end with a cork was filled with distilled water heated to about 50°C. The mouth of the cylinder was closed with a cork through which a thermometer was inserted. To simulate the actual condition, a wire mesh has been wrapped on the surface of the cylinder to obtain a clearance of 2 mm between fabric sample and brass cylinder. A rectangular specimen of the fabric was used to cover the entire outer surface of the brass tube. The length-wise edges of the specimen were made to touch each other closely avoiding overlapping and kept in position by using cello-tape over the joint running parallel to the length of the cylinder [3].

The experiment was started when the temperature of the water was exactly 48°C. A stopwatch was used to find the time taken for the temperature fall at every 1°C. A cooling curve was drawn from these data and the time taken to cool from 48°C to 38°C was found. The TIV was calculated by using Marsh method, as follows [3, 5]:

$$\text{TIV} = \left[1 - \frac{t_c}{t_u}\right] \times 100 \quad (3)$$

where, (t_c) is time taken by the covered body to cool through a certain temperature range and (t_u) is time taken by the uncovered body to cool through the same temperature range. They found that TIV is related to the thickness of the fabric, the basis weight (fabric weight), and the number of layers of the fabric [1]. The intra fabric air spaces and inter space between fabric and body are also important. The TIV of the fabric is greater when a non-conducting mesh (polythene) is present between the cylinder and fabric instead of conducting metal mesh in the same position. Increase in any of these factors increases the TIV significantly. There has been marginal effect on TIV with varying fabric nature.

3.6. Thermal insulation behaviour of jute-based knitted fabrics

Fabric structure plays a very important role in thermal insulation property, which has been mentioned earlier. Further in the same line, Vigneswaran et al. have studied knitted jute-based fabric structure [15]. They studied the effect of thermal conductivity behavior of jute/cotton blended knitted fabrics. The thermal conductivity is the reciprocal of the thermal insulation. They established the relationship between fabric properties and thermal conductivity of various developed jute/cotton blended knitted fabrics. Experimental result obtained by them confirms that lower thermal conductivity is obtained at higher jute blend proportions. They concluded that the thermal conductivity reduces with increasing fabric thickness. This study also reveals that fabric air permeability and fabric tightness factor values influences the thermal conductivity of jute/cotton blended knitted fabrics. Higher TIVs are noticed with higher fabric tightness factor and lower air permeability [15]. Regression correlation coefficients between various fabric properties and thermal conductivity have also been discussed.

3.7. Thermal insulation behaviour of jute-based warm garments

It has been proven on evidences from the literature that jute-based fabrics are having equally good thermal insulation property when compared with synthetic acrylic and cotton shawl materials [11]. Jute and hollow polyester materials are used to prepare the weft yarns of the shawl and cotton yarn were used in warp direction to weave the shawl fabric. Apart from its thermal insulation property, other properties like air permeability, fabric cover factor is also better in case of the developed jute-polyester and cotton blended shawls. Furthermore, jute and polyester and cotton blended union fabric have been used to develop the jacket for winter season [8, 12, 13]. It has been found from this study that the jackets are comparable or better compared to commercial polyester jackets of the same jacket weight [9].

4. Conclusions and future prospects

It can be concluded from this study that jute-based material can be used effectively in different thermal insulating application. These applications are shawl, jacket, blanket, carpet, etc. Jute-based materials also have immense potential in other industrial applications as thermal insulation material.

Apart from these jute-based thermal insulation materials, electromagnetic shielding resistance, vibration resistance/insulation material, mechanical shock resistance/insulation, electrical insulation material, sound/noise insulation material, etc. are the future research directions. There are enormous scopes as insulation material from jute-based material for different domestic, industrial, and apparel applications. Jute-based textile as insulation material may be considered as green/sustainable material, which may replace a greater extent of synthetic material for the same application. Finally, it can be concluded that jute and jute-based allied fibres will get new avenue in future days as far as application of insulation material is concerned.

Author details

Sanjoy Debnath

Address all correspondence to: sanjoydebnath@yahoo.com

Division of Mechanical Processing, ICAR-National Institute of Research on Jute & Allied Fibre Technology, Kolkata, West Bengal, India

References

[1] Debnath C. R., Bhowmick B. B., Ghosh S. K., and Das P. K. Thermal insulation behaviour of some nonwovens. Textile Trends, 1987:30(5):45–49.

[2] Debnath S., Ganguly P. K., De S. S., and Nag D. Control of soil moisture and temperature by light weight jute fabrics. Journal of the Institution of Engineers (India), 2010:90TX(2): 16–19.

[3] Debnath S., Kane C. D., Kadole P. V., and Patil C. A. Needle-punched non-woven blankets from polyesters. Indian Textile Journal, 1994:105(3):72–80.

[4] Debnath S., and Madhusoothanan M. Thermal resistance and air permeability of jute-polypropylene blended needle-punched nonwoven. Indian Journal of Fibre & Textile Research, 2011a:36(2):122–131.

[5] Debnath S., and Madhusoothanan M. Thermal insulation, compression and air permeability of polyester needle-punched nonwoven. Indian Journal of Fibre & Textile Research, 2010:35(1):38–44.

[6] Debnath S., and Madhusoothanan M. Thermal resistance behaviour of polyester needle-punched nonwoven. Journal of the Institution of Engineers (India), 2011b: 91TX(2):27–33.

[7] Booth J. E. *Principles of Textile Testing*. 3rd ed. London: Newnes Butterworth; 1976. pp. 100–101.

[8] Anonymous. Annual Report 2007–2008. National Institute of Research on Jute & Applied Fibre Technology, ICAR, 12, Kolkata, India; 2008. p. 33.

[9] Yachmenev V. G., Parikh D. V., and Calamari T. A. Thermal insulation properties of biodegradable, cellulosic-based nonwoven composites for automotive application. Journal of Industrial Textiles, 2002:31(4):283–296.

[10] Debnath S., Bhattacharya G. K., and Singh U. S. A blanket from jute-hollow polyester blended bulk yarn and method of preparing the same. Indian Patent Application No. 1102/KOL/2009, August 28, 2009.

[11] Debnath S., Sengupta S., and Singh U. S. A method for producing jute-hollow polyester blended yarn, union fabric of said yarn and method of preparing the said union fabric and shawl from the said yarn. Indian Patent Application No. 1187/KOL/2008, July 9, 2008.

[12] Anonymous. Annual Report 2005–2006. National Institute of Research on Jute & Applied Fibre Technology, ICAR, 12, Kolkata, India; 2006. pp. 33–35.

[13] Anonymous. Annual Report 2006–2007. National Institute of Research on Jute & Applied Fibre Technology, ICAR, 12, Kolkata, India; 2007. pp. 29–30.

[14] Anonymous. Annual Report 2010–2011. National Institute of Research on Jute & Applied Fibre Technology, ICAR, 12, Kolkata, India; 2011. pp. 17–18.

[15] Vigneswaran C., Chandrasekaran K., and Senthilkumar P. Effect of thermal conductivity behavior of jute/cotton blended knitted fabrics. Journal of Industrial Textiles, 2009:38(4):289–307.

Section 2

Practical Applying and Performing of Insulation Materials (Case Studies)

Chapter 4

Seismic Aspects of the Application of Thermal Insulation Boards Beneath the Foundations of Buildings

David Koren, Vojko Kilar and Boris Azinović

Additional information is available at the end of the chapter

http://dx.doi.org/10.5772/62294

Abstract

In recent years, there has been a significant increase in the construction of energy-efficient buildings. These buildings are mainly characterized by their thermal envelope, which needs to follow the complete outer perimeter of the building without any interruptions, to avoid thermal bridges. It has been observed, however, that the specific new details which prevent the occurrence of thermal bridges can, in many cases, substantially affect the structural integrity of such buildings during earthquakes. This chapter deals with the seismic aspects of the application of thermal insulation (TI) boards beneath the foundations of buildings. For this purpose, the mechanical characteristics of the most commonly used material in practice (i.e., extruded polystyrene — XPS) were experimentally determined. Additionally, the shear behaviour of differently composed TI foundation sets was investigated and their friction capacity estimated. The authors have proposed a new solution for the foundation detail, which is based on controlling the sliding mechanism between the individual layers of TI boards in order to reduce the seismic forces induced on the superstructure. The proposed detail with a specially designed sliding layer surface is made of commonly used TI materials for modern passive houses, thus reducing the potential additional costs. The solution was verified by means of nonlinear dynamic analysis of several realistic building models and various friction coefficients between XPS boards. The selected results are presented in terms of fragility curves for the occurrence of sliding between the layers of XPS boards. Based on these curves, the desired seismic response scenario and level of protection of a building structure could be selected.

Keywords: thermal insulation, foundations, seismic response, energy-efficient building, extruded polystyrene (XPS), friction, sliding isolation

1. Introduction

In the European Union (EU), a large part of total energy consumption is due to the high demands for the heating/cooling of the existing building stock. It has been estimated that buildings alone cause 40% of the total energy consumption in the EU [1]. Reducing energy consumption and producing energy from renewable sources therefore represent an important research issue in the building sector. Consequently, over the last 15 years there has been an increasing trend towards the construction of energy-efficient buildings, which was also stimulated by the implementation of the directive 2010/31/EU [1]. New requirements for the energy efficiency of buildings have been set, and the year 2020 has been defined as a milestone for nearly zero-energy buildings. However, the earthquake safety of energy-efficient buildings, as well as their new construction details, needs to be investigated. In those parts of Europe where the construction of low-energy buildings has already become an established practice, earthquakes are for the most part unknown, so that such verification of new construction details is not necessary. In recent years, however, low-energy building standards [2–5] have been slowly gaining ground in areas where earthquakes (including strong earthquakes) are frequent, such as Spain, Portugal, Italy, Greece, Croatia, Slovenia and others. The suitability of the newly developed construction details for energy-efficient buildings therefore needs to be verified, and appropriate solutions found also for earthquake prone regions. This chapter emphasises on the investigation of a newly proposed structural detail which would on the one hand prevent the formation of thermal bridges, while on the other hand, it would reduce possible damage of the superstructure by acting as a seismic protection fuse during strong earthquakes.

2. Materials for thermal insulation boards

The most frequently used technique for preventing a thermal bridge occurring beneath the ground floor slab is founding of the building on thermal insulation (TI) boards, which need to have adequate compressive strength. Extruded polystyrene (XPS) boards, expanded polystyrene (EPS) boards and boards made of cellular glass are most frequently used for this purpose. They are usually placed under the reinforced concrete (RC) foundation slab or under the strip foundations in one or more layers. In order that the insulation material remains thermally functional during the whole of its expected life time, it should be resistant to long-term loadings, temperature changes (i.e., freezing/thawing cycles), creep, shrinkage and all the effects of frost, moisture and ground water. The most preferred type of boards are those whose thermal and physical characteristics do not change even if they are in constant contact with water. An example of the casting of a RC foundation slab on top of layers of XPS insulation boards is presented in **Figure 1**.

http://dx.doi.org/10.5772/62294

Figure 1. Example of extruded polystyrene (XPS) insulation boards used under a foundation slab [6].

Material property	Cellular glass	Extruded polystyrene (XPS)	Expanded polystyrene (EPS)	Polyurethane (PUR/PIR)	Mineral wool
Density ρ [kg/m^3]	100–165	25–35	15–30	30–100	40–200
Thermal conductivity λ_D [W/mK]	0.040–0.065	0.030–0.040	0.031–0.043	0.020–0.035	0.03–0.045
Water absorption W_{lp} [volume %]	<0.2	<0.3	<1.0	<1.6	<3.0
Compressive strength σ_{10} [kPa]	400–1600	100–1000	30–500	25–800	10–90
Compressive creep strength σ_{cc} [kPa]	100–700	20–300	10–150	5–250	2–30
Elastic modulus E [MPa]	100–500	15–40	5–25	2–25	0.3–2.0
Shear strength τ [kPa]	80–400	100–200	10–300	100–450	5–50
Shear modulus G [MPa]	>4.0	3.0–8.0	1.5–9.0	1.0–5.0	0.3–1.5
Energy for production* [kWh]	85	43–89	39–95	47–64	9–90
Relative material cost*	5.3–5.9	3.0–3.5	1.0–1.2	>3.0	1.0–1.5

*Energy for production and relative cost are calculated for the same thermal conductivity value of U = 0.4W/m^{2}K for both materials. The price of EPS is used as a unit [19].

Table 1. Properties of thermal insulation boards used for below-grade applications.

The most important mechanical and TI properties of commonly used TI boards, as can be found in recent catalogues and other available data in the literature [6–16], are presented in **Table 1**. The values presented in **Table 1** were obtained in monotonic static tests, whereas some of the results of cyclic tests needed for earthquake simulations are presented in Section 3. Protocols for the testing of materials and the obtaining of appropriate EU certificates are prescribed in the corresponding EU standards. For example, the presented values for compressive strengths were obtained according to the standards EN 826 [17] and EN 1606 [18], where the maximum compressive strength σ_{10} is defined at a deformation of 10%, and the long-term maximum compressive creep strength for an assumed building life-time of 50 years (σ_{cc}) is defined as the strength corresponding to a deformation of 2%.

It can be seen from **Table 1** that boards made of mineral wool have a very low compressive strength and elastic modulus, so they are presumably not suitable as TI under the foundations of a building. As well as this, mineral wool is not suitable because it can absorb more water than other materials. Polyurethane, too, appears to be unsuitable because of its lower thermal conductivity and also because the production of polyurethane is related to environmental threats to the Earth's ozone earth layer. Thus, three materials remain as being suitable for TI under foundations: XPS, EPS and cellular glass. Their microstructures are shown in **Figure 2**.

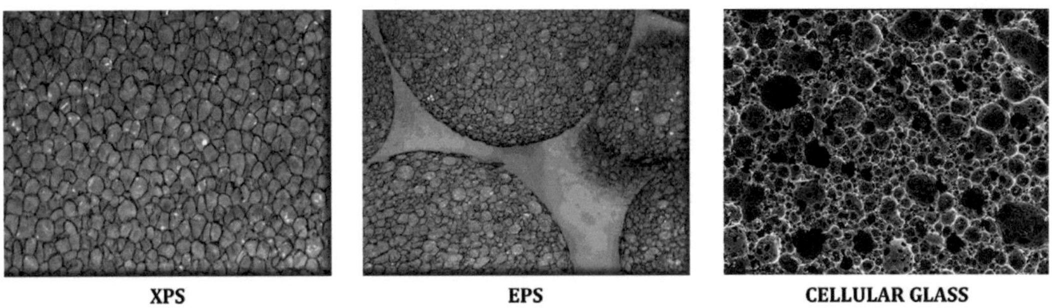

Figure 2. The microstructure of thermal insulation materials, magnified 25 times [20, 21].

XPS and EPS boards are both made from polystyrene, but their production processes (expansion or extrusion) are different. They are both formed of round-shaped grains, which are in the case of EPS polyhedral in shape and held together by process of expansion [8, 13]. The extrusion process used to produce XPS results in a more homogenous structure, which is also more repellent to water. This means that such material is more suitable in all structural details where the presence of moisture can be expected. Research into the mechanical characteristics of EPS has shown that it maintains its strength and stiffness also if it is in constant contact with water [22–24]. However, it has been shown that its thermal conductivity might increase if it is not additionally protected by a waterproofing layer. In comparison with EPS and XPS, cellular glass is more similar to XPS, since it also has a closed cell structure and is therefore highly repellent to water. Among all these three materials, EPS has the lowest strength and stiffness, but it remains interesting due to the much lower costs of its production. EPS can also be premoulded into various blocks and modules, which can replace formwork and therefore significantly speed up the construction of concrete buildings. Some house manufacturers [13] allow EPS to be used under the foundation slabs of one or two storey RC/masonry houses, or even more storeys in the case of houses made of lighter materials (e.g., wood). For higher/slenderer buildings, the use of XPS or cellular glass is more appropriate. Among these, cellular glass has the best strength and stiffness characteristics, but it is much more expensive to produce and less thermally efficient. For these reasons, it is most frequently used for the insulation of multi-storey or more complex structures, where higher strength is needed.

Presently XPS is the most frequently used material for TI under foundations. It is based on the use of a polymerised polystyrol and a foaming agent. In general, rather limited research has been performed up until now into the behaviour of XPS foam, although the behaviour of this foam under monotonic compressive loading conditions is regularly controlled during the production process. In the relevant scientific literature, only a few references can be found in relation to the behaviour of XPS foam. Improved XPS foam insulation with better material efficiency (and lower thermal conductivity) was developed in [9]. An experimental study concerning the hygrothermal behaviour of retrofit solutions as applied to older stone masonry walls has, for instance, been presented in [25]. In [26] XPS foam was applied as part of a vibration isolating screen installed in soil near a test public transport track. In this research some dynamic material characteristics of the used XPS were determined by means of a white noise–forced vibration test on a freely suspended bar element: it was found that the dynamic Young's modulus of the XPS was equal to 35 MPa, whereas its density was equal to 45 kg/m^3, its Poisson's ratio to 0.2 and its material damping to 1.0%. The long-term mechanical properties (i.e., compressive creep strains and modulus), which are of key importance for TI placed under foundations, have been analysed in [27]. In the same reference, the modelling of a foundation slab resting on a TI layer has also been schematically indicated. Vaitkus et al. [28] experimentally analysed XPS short-term compression dependence on exposure time. Significant changes in the XPS strength characteristics after 45 days were observed. The relationship between the XPS foam microstructure and its response under compressive load has been analysed in [29]. The average compressive strengths of the tested samples were equal to 729, 347 and 179 kPa for the normal, transverse and longitudinal directions, respectively, and the corresponding moduli of elasticity amounted to approximately 37.0, 16.7 and 5.7 MPa. Morphological data about the XPS boards were obtained by using the X-ray tomography imaging technique and then used to develop microstructure-based finite element models. In the parametric study the effect of cell size and cell anisotropy on the mechanical response of XPS boards under compressive loads was analysed. It was shown that the microstructure cell size has no effect on the mechanical properties of XPS rigid boards when loaded in compression as long as the density of these boards remains constant. On the other hand, the degree of cell anisotropy was found to have a very important influence. Similar findings about the influence of the polymeric foam's density on mechanical properties in tension and compression have been reported in [7], where the study concludes that the foam's (e.g., XPS) deformation pattern beyond the yield point in compression is non-homogeneous. A structural macro- and microstructure analysis of XPS foam with terahertz spectroscopy and imaging was performed in [30].

As far as is known, the cyclic compressive stress–strain behaviour of XPS foam, which is essential for its seismic response in earthquake engineering applications, has not yet been researched. Furthermore, information about the cyclic behaviour of this material when subjected to shear loadings is currently a completely unresearched issue. From this point of view, extensive experimental research addressing the cyclic behaviour of XPS foam in compression, as well as in shear, has been recently performed in order to obtain a better understanding of the fundamental behaviour of this foam in earthquake engineering applications [31]. Laboratory tests were carried out on two different XPS products. These were XPS boards that were manufactured by a Slovenian enterprise and denoted as 400 L and 700 L (the number in the denotation indicates the nominal compressive strength of the XPS material in kPa). The product data are given in [6]. A servo-hydraulic testing machine was used for these tests, whereas the deformations of the XPS specimen were monitored by means of LVDTs. In the next section, the main results of the performed laboratory tests are summarized.

3. Laboratory tests

3.1. The behaviour of the XPS boards under compressive load

The compressive behaviour of the investigated XPS products was determined according to the standard SIST EN 826:1997 [17], with some modifications. The details can be found in [31]. During the performance of the monotonic compressive tests, the distance between the fixed and the movable steel plates of the testing machine was reduced until the relative deformation of the specimen reached a final value which was close to 90% (**Figure 3**). However, the compressive stress at a relative deformation of 10% was still used to estimate the XPS product's compressive strength, according to the provisions of the corresponding standard [17]. The aim of the measurements beyond a relative deformation of 10% was to determine the relative deformations of the XPS specimens, where unloading would start in the case of cyclic compressive tests. In these tests the testing procedure was similar to that used in the monotonic tests, but with additional unloading-reloading cycles at 20%, 40%, 60% and the final relative deformation of the XPS specimen.

The results of the compressive monotonic and cyclic tests are presented in **Figure 4**. The estimated compressive strengths (σ_{10}) and the compressive moduli of elasticity (E), calculated as the average of all the test results (both monotonic and cyclic), are given in **Table 2**, along with the corresponding coefficients of variation (COV). The obtained results show that the behaviour of the XPS products in compression is characterized by three regions – elastic, plastic and densification, which is typical behaviour for foamed products [28, 32–34]. The average experimentally obtained compressive strengths (σ_{10}) of the 400-L and 700-L products were 490 kPa and 752 kPa, respectively, which means that they were 22.5% and 7.4% greater than the declared compressive strength of the two products, respectively. The response of the 400-L product, which had a lower density due to the lower content of polystyrene than in the case of the 700-L product, seems to be a combination of the behaviour of elastomeric foams up to a compressive deformation of about 10%.

http://dx.doi.org/10.5772/62294

By comparing the response of the XPS products subjected to the monotonic and cyclic compressive tests, it can be seen that stress-strain envelope in the case of cyclic response corresponds to the stress-strain response obtained in the monotonic test, for both the 400-L and the 700-L products. When unloading the specimen at relative deformations of 20%, 40% and 60%, the so obtained diagram is first parallel to the elastic region, whereas later on the tangent *E*-modulus starts to decrease until a residual plastic deformation is reached at zero stress.

Figure 3. XPS specimen before and at the end of the compressive test.

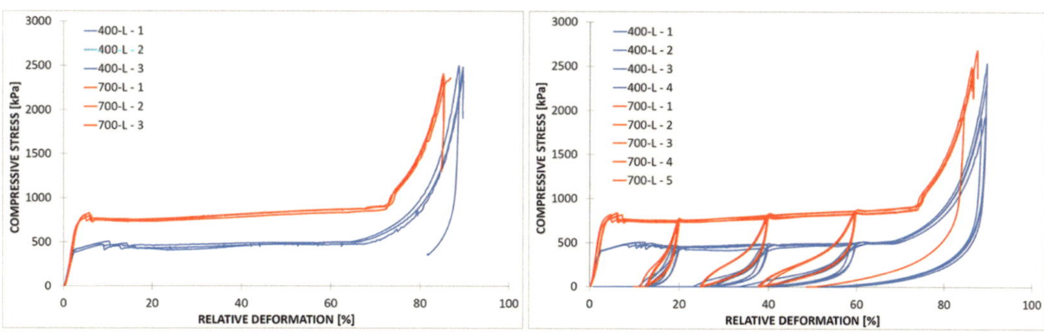

Figure 4. Compressive stress–relative deformation diagrams of the 400-L and 700-L specimens in the case of monotonic (left) and cyclic (right) compressive tests.

3.2. The behaviour of the XPS boards under shear load

The shear behaviour of the XPS products was determined according to the standard SIST EN 12090:1999 [35], using a double test specimen assembly. The test specimen consisted of two XPS blocks which were glued onto a system of three parallel steel plates where the middle plate was movable and the outer plates were fixed (**Figure 5**). The details can be found in [31]. When the monotonic shear test was carried out, the standard procedure was followed until failure of the specimen, and the shear deformation at failure was used as a reference value in order to determine the deformations where unloading of the specimen started in the case of a cyclic shear test, in which some modifications of the standard procedure were introduced. One cycle of the shear test consisted of loading until the selected deformation was reached in the "+" direction, followed by unloading to zero load, reloading until the same absolute value of deformation was obtained in the "−" direction, and then unloading to zero load. The selected deformations where unloading of the test specimen began for the cyclic tests were equal to 20%, 40%, 60%, 80% and finally 100% of the reference deformation.

Figure 5. The XPS shear test setup.

Selected results of the shear tests are presented in **Figure 6**. The measured shear strengths (τ) and shear moduli (G), calculated as the average of all the test results (monotonic and cyclic in the "+" direction), are given in **Table 2**, together with the corresponding values of the coefficient of variation. From the results presented in **Table 2** it can be concluded that the shear strength of the 400-L and 700-L products is 3.5 times less than their corresponding compressive strength, whereas the G-modulus of the products is about 5 times less than the corresponding E-modulus. Comparing the response of the XPS in compression and in shear, it can be seen that the obtained shear ductility capacity is smaller, and that strength degradation is evident in the deep nonlinear behaviour range. By comparing the response of the XPS products obtained in the monotonic and cyclic shear tests (**Figure 6**) it can be seen that the hysteresis envelope in the case of cyclic response in the "+" direction corresponds to the stress-strain response obtained in the monotonic test, for the 400-L and 700-L products. For both products the deformations corresponding to the shear failure of the foam are about the same (between 9 and 11%), but the shear strength is about 50% higher in the case of the 700-L product. Thus, the 700-L foam is able to absorb more energy in shear, too. A similar finding has been made in the case of EPS material [36]. Observing the cyclic behaviour of the 400-L and 700-L products it can be concluded that unloading at 20% of the reference deformation occurred in the elastic region, leading to zero or negligible residual plastic deformation in the tested material. When the unloading deformation was increased to 40%, 60%, 80% and 100% of the reference shear deformation, the residual plastic deformation gradually increased, too. It should also be noted that the hysteresis is not symmetric – the symmetry axis is shifted slightly into the first quadrant of the coordinate system.

	COMPRESSIVE TESTS		**SHEAR TESTS**	
Product	σ_{10} [kPa]/COV [%]	E [kPa]/COV [%]	τ [kPa]/COV [%]	G [kPa]/COV [%]
XPS400-L	490/3.5	24200/5.8	138/1.7	4520/5.1
XPS700-L	752/1.1	36100/5.3	219/5.3	7460/2.6

Table 2. Average values of the compressive strength σ_{10} and the modulus of elasticity *E*, and of the shear stress τ and shear modulus *G*.

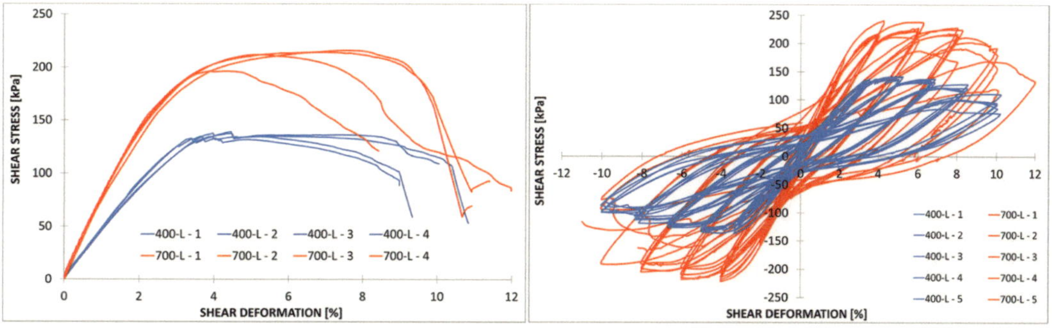

Figure 6. Shear stress–relative deformation diagrams obtained for the XPS specimens in the case of monotonic (left) and cyclic (right) shear loading tests.

3.3. Shear behaviour of the analysed TI foundation sets

In order to estimate the coefficients of friction between the different constituent elements in the TI foundation set, various shear tests were carried out. These tests have not yet been standardised and were specially developed for the needs of our experiments [31, 37]. Various TI foundation sets were analysed (**Figure 7**) – consisting of one or two XPS boards, a concrete slab, with/without a waterproofing insulation (HI) or a polyethylene (PE) sheet. At a selected level of pre-compression (from 50 to 300 kPa) in the vertical direction of the tested set, horizontal displacements were induced by means of a servo-hydraulic actuator. The levels of pre-compression were selected based on the likely levels of compressive stress beneath the foundation slab during a moderate earthquake. For each tested TI foundation set the response and the coefficients of static (μ_s) and kinetic (μ_k) friction were determined.

Scheme of TI foundation set	Specimen ID*	Description
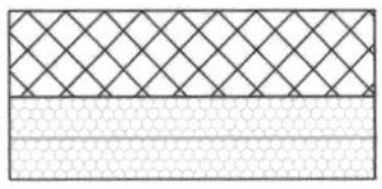	1-XPS	2 x 120 mm XPS 400-L, without a PE or HI sheet
	2-XPS	2 x 100 mm XPS 700-L, without a PE or HI sheet
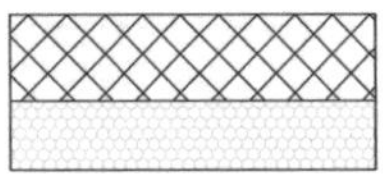	3-XPS	1 x 200 mm XPS 400-L, without a PE or HI sheet
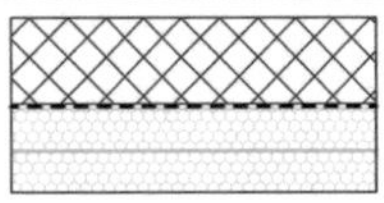	4-XPS	2 x 120 mm XPS 400-L, with a PE sheet
	5-XPS	2 x 100 mm XPS 700-L, with a PE sheet
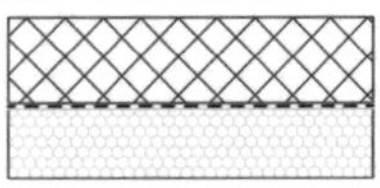	6-XPS	1 x 200 mm XPS 400-L, with a PE sheet
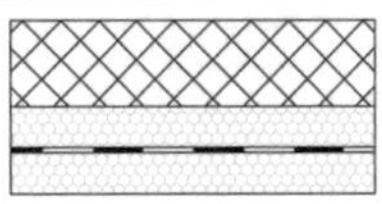	7-XPS	2 x 120 mm XPS 400-L, HI sheet without adhesive
	8-XPS	2 x 100 mm XPS 700-L, HI sheet without adhesive
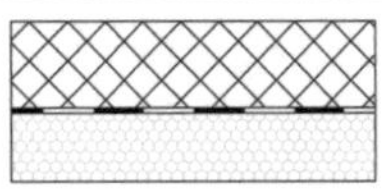	9-XPS	1 x 200 mm XPS 400-L, HI sheet without adhesive
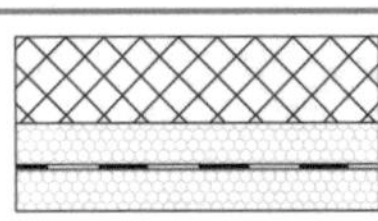	10-XPS	2 x 120 mm XPS 400-L, HI sheet having adhesive on one side only
	11-XPS	2 x 100 mm XPS 700-L, HI sheet having adhesive on one side only
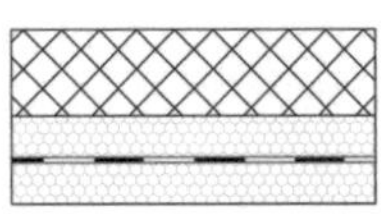	12-XPS	2 x 120 mm XPS 400-L, HI sheet having adhesive on both sides
	13-XPS	2 x 100 mm XPS 700-L, HI sheet having adhesive on both sides

*** In further text and figures the test specimen's ID designation is followed by the number (50–300) representing the level of applied pre-compression level in [kPa].**

Figure 7. The tested TI foundation sets.

The tests have showed that, in the case of sliding, the XPS–HI sheet contact (without adhesive on one/both sides) which was exposed to lower levels of pre-compression (e.g., 50 kPa) was the critical one. The corresponding coefficient of friction was around 0.3 [31]. In practice this means that, during a strong earthquake, it is possible that a passive house with such a foundation set would slide in the horizontal direction (**Figure 8**).

http://dx.doi.org/10.5772/62294

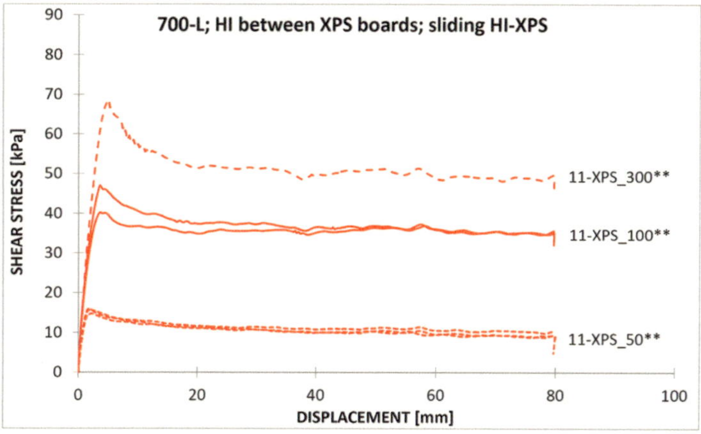

Figure 8. The 700-L TI foundation set with a HI sheet having adhesive on one side only between two XPS boards: stress–displacement diagrams at different levels of pre-compression (left) and sliding at a pre-compression level of 100 kPa (right) [31].

Another tested TI foundation sets (1-XPS and 2-XPS) consisted of two XPS boards which were in contact with the levelling concrete (without a HI or PE sheet being inserted between them). The two contact planes were carefully monitored: the XPS–XPS contact, and the XPS–levelling concrete contact. The results obtained in the tests showed that sliding occurred between the two XPS boards before sliding between the XPS and the levelling concrete. However, the test of a TI foundation set consisting of two 120 mm thick XPS boards (400-L) and levelling concrete showed a higher sliding resistance than the corresponding TI foundation set 3-XPS, which consisted of one thicker (200 mm) XPS board. The resulting coefficients of friction are presented for both TI foundation sets in **Table 3**. In the same table the obtained coefficients of friction for the TI foundation sets 4-XPS and 6-XPS are shown. These foundation sets consisted of one or two 400-L XPS boards, levelling concrete and a PE sheet inserted between them. In this case, at low pre-compression levels, sliding was observed at the contact between the PE sheet and the XPS layer or levelling concrete. At a higher pre-compression level (300 kPa) deformation of the XPS boards was observed only in the case of the two XPS boards. On the other hand, in the case of the TI foundation set with one thicker (200 mm) 400-L board (6-XPS) exposed to a pre-compression level of 300 kPa, a combination of deformation of the XPS boards and sliding between the PE sheet and the XPS was the typically observed response.

It can be concluded that TI foundation sets without a PE sheet between the levelling concrete and the XPS in general provide higher shear capacity than corresponding foundation sets with an inserted PE sheet. In the case of the TI foundation set with two 400-L boards (2 x 120 mm) this increase amounted to about 50%. The same values as those given in **Table 3** for the 2 x 120 mm 400-L boards (1-XPS, 4-XPS) can also be used in the case of the 2 x 100 mm 700-L boards (2-XPS, 5-XPS) since the μ_s values are only slightly higher in the case of the foam with a higher density.

	TI foundation set without a PE sheet between the levelling concrete and the XPS board(s)		TI foundation set with a PE sheet between the levelling concrete and the XPS board(s)	
Number and thickness of 400-L XPS boards in the TI foundation set	μ_s	μ_k	μ_s	μ_k
2 × 120 mm	0.600	0.570	0.413	0.365
1 × 200 mm	0.487	0.459	0.520	0.414

Table 3. Average values of the coefficient of maximum static (μ_s) and kinetic (μ_k) friction for the 1-XPS, 3-XPS, 4-XPS and 6-XPS foundation sets at a pre-compression level of 50 kPa.

4. A new technological solution for the foundation of passive houses in seismic areas

4.1. Description of the proposed foundation system

Recently, in [38], a new solution has been proposed with regard to the design of foundations for thermally insulated passive houses. The proposed solution is based on controlling the sliding between the individual layers of XPS boards in order to reduce the seismic forces which are induced on the superstructure. It is protected by a patent that has been filed at the Slovenian Intellectual Property Office (SIPO). The principle of the solution is analogous to that of sliding seismic base isolation systems [39–47].

The solution has been developed based on existing passive house foundation details, which were designed in order to prevent the occurrence of thermal bridges running from the heated interior of the building to the ground underneath. The proposed solution still permits the use of existing foundation construction details, while its added value consists of the additional components for the controlled response of buildings in seismically active areas, taking advantage of the sliding effect. The additional components are shown in **Figure 9** and are marked as follows; (1) vertical restrainers for the prevention of uncontrolled rocking and larger lateral shifts, (2) a lateral sliding gap (ΔH), (3) the imposed sliding surface and (4) horizontal stoppers for the prevention of sliding at the contact surface between the blinding concrete and the first layer of TI. The red arrows shown in Section A-A of **Figure 9** indicate the possible movement of the building during earthquake ground motion — if the upper part of the foundation detail shifts, together with the building, to the left, then the size of the lateral sliding gap (ΔH) on the left hand side of the building will increase and the sliding gap on the right hand side will decrease (until blocked by horizontal stoppers). In the case of very large horizontal shifts, the sliding displacement is limited by the size of the lateral sliding gap (ΔH). In this case the vertical restrainers collide with the horizontal stopper on one side of the foundation slab, whereas they separate on the opposite side until a maximum gap opening size of $2 \cdot \Delta H$ is reached. It should be mentioned that the size of the lateral sliding gap (ΔH) is defined by the buildings' designer, who can therefore limit the size of the maximum residual displacement. On the other hand, the vertical restrainers prevent uncontrolled rocking of the building, as well as the occurrence of any irreversible compressive deformations of the TI layer(s). The vertical restrainers are located at defined distances around the foundation slab and are separated from one another by the distance ΔR (**Figure 9**). Vertical restrainers and horizontal stoppers can, in practice, be produced in many different ways – depending on the decision of the designer/investor and the availability/prices of the available components. It is, however, necessary, that the vertical restrainers are made of material with a high compressive strength and simultaneously a low thermal conductivity. They can consist of various hollow steel sections filled with TI, or of special pressure bearings made of thermal insulating (nano) concrete, or even of solid RC edge elements, which are additionally thermally insulated. At the contact with the soil, the vertical restrainer ends with a flat and wide surface, which prevents the negative effect of the restrainer penetrating into the soil due to the applied concentrated load (the restrainers must be placed close enough to each other — ΔR). Horizontal stoppers need to provide sufficient shear stiffness and shear strength. It is also recommended that the perimeter of the foundation base (e.g., blinding concrete) should be strengthened in order to increase the shear capacity of horizontal stoppers and to prevent protrusion of vertical restrainers.

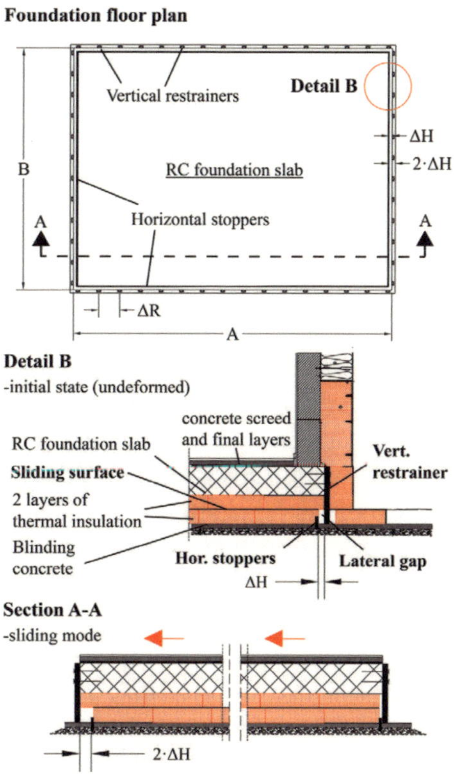

Figure 9. Section A-A and the foundation slab layout with the included anti-seismic sliding components.

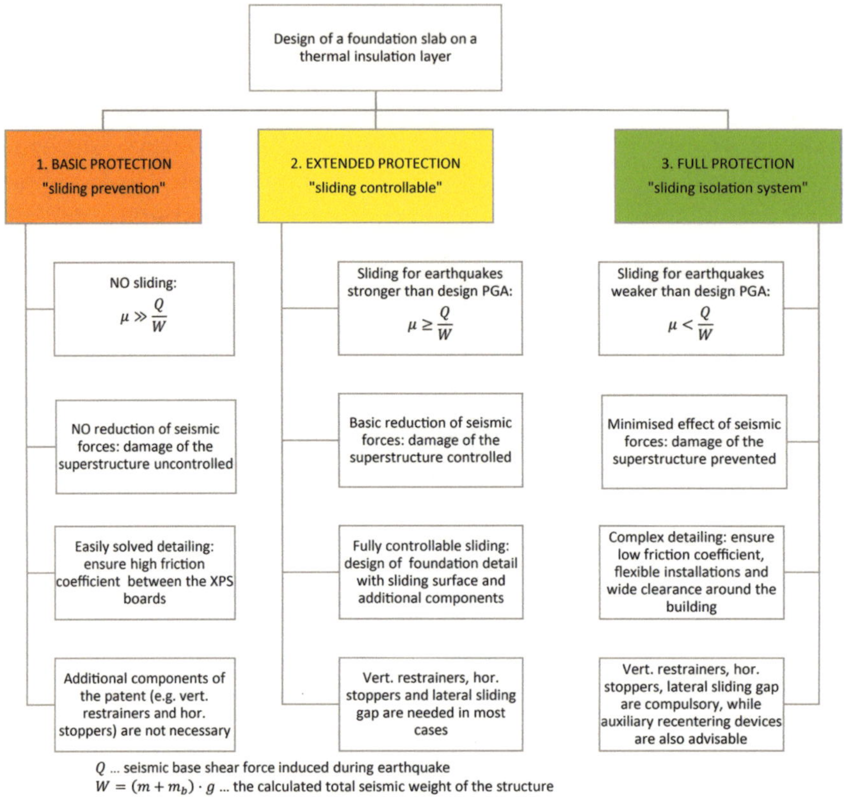

Figure 10. Overview of the characteristics of the possible seismic response scenarios.

The design of passive buildings could, according to the proposed solution, follow any of the three different seismic response scenarios that are presented in **Figure 10**. The first scenario avoids sliding and its potential benefits, but ensures basic protection of the superstructure. This concept is mainly used for simple buildings (e.g., 1–2 storey buildings), which are not vulnerable to strong seismic shaking. For such buildings the use of a sliding layer with a higher friction coefficient is advisable, since the seismic forces do not need to be reduced. In this case the application of the additional components of the proposed foundation detail (vertical restrainers, horizontal stoppers and a sliding gap) is unnecessary. The main concern, in this case, is only to ensure a sufficiently high friction coefficient between the individual layers of the foundation detail, so that the sliding mechanism is eliminated and the building's installations remain intact as in the case of conventionally founded (fixed base) buildings. For this reason, this scenario is referred to as "basic protection" (or "sliding prevention"). In the case of the second and third scenarios a reduction in the seismic forces acting on the superstructure can be achieved. The difference between the second scenario (referred to as: "extended protection" or "sliding controllable") and the third scenario (referred to as "full protection" or "sliding isolation system") depends on the seismic intensity which will activate the sliding mechanism. In the second scenario the sliding mechanism is activated only in the case of earthquakes with seismic intensities higher than the design seismic intensity, whereas in the case of the third scenario sliding can also occur in the case of weaker earthquakes. The building designer and investor can therefore, together, choose the desired level of protection of the building (basic, extended, or full protection). In the case of third scenario the friction coefficient needs to have a much lower value, so that it is comparable with the friction coefficients of sliding isolators. It should be noted that, in practice, it is difficult to achieve such a low friction coefficient in the case of large areas of the foundation slab detail, so that additional costs would probably be incurred. Furthermore, in the third scenario the building designer needs to provide all of the additional components of the proposed solution which are necessary for it to be effective, i.e., flexible installations, horizontal stoppers, a lateral sliding gap (ΔH) and sufficiently wide clearance around the building. The latter is particularly important in order to avoid any pounding of an isolated structure against adjacent buildings [47].

For all of the above stated reasons the authors believe that, in the case of modern passive houses, it would be best, for the time being, to recommend the second scenario. The latter makes use of commonly available materials that have already been used in existing passive house foundation details, so that no additional costs would be incurred if a seismic fuse were to be created which would be activated only when a strong earthquake occurs.

4.2. Numerical verification and selection of the seismic response scenarios

The applicability of the proposed solution was demonstrated by means of the nonlinear dynamic analysis of some simplified parametric models, as well as of some realistic models of two, four and six storey RC passive house buildings. An extensive description of the performed analyses and the results obtained can be found in [38], whereas in this chapter only selected results are presented.

4.2.1. SDOF superstructure model

In the first stage of the study the effectiveness of the proposed solution was analysed by using a simplified Single Degree-of-Freedom (SDOF) model (**Figure 11**) which represents 2-storey (H = 6 m) passive house structures with short fundamental periods (T_{FB} = 0.10 – 0.30 s). The structures were assumed to be founded on a foundation slab (with floor plan dimensions A/B = 16/8 m) and supported by nonlinear springs in order to model the XPS boards with a nominal strength of 400 kPa and a total thickness of d = 30 cm (2 layers of thickness 15 cm). The horizontal spring for the sliding resistance of the TI foundation detail was defined based on the initial vertical compressive stresses due to the assumed mass of the superstructure during an earthquake ($m + m_b$). Six different friction coefficients (TI foundation sets) were investigated. Two different models of the superstructure were analysed, one showing elastic behaviour (with a strength factor α = 1.0 and a ductility capacity δ_c = 1.0) and the other showing nonlinear behaviour (α = 0.25 and δ_c = 4.0). The models showing elastic behaviour were investigated in order to determine the maximum response of the foundation demand parameters such as the base displacement (D_{base}), whereas the nonlinear superstructure models were used for comparison of the superstructure demand parameters, such as the ductility demand (δ_d). It should be noted that such a selection was made according to [48, 49], where it was shown that elastic models yield maximum base engineering demand parameters (EDPs), whereas inelastic models yield maximum superstructure EDPs. The seismic response of the structural models was evaluated by means of nonlinear time-history analyses considering a set of 30 real ground motion records (GMRs), which were selected so that they matched the target spectrum for stiff soil sites (soil type A in EC8 [50]), with 5% damping and a seismic intensity of 0.25 g. Incremental dynamic analyses (IDA) were performed for each selected GMR, with a variation step of 0.02 g up to a maximum seismic intensity of 1.0 g. Uni-directional dynamic analysis was adopted, which was best suited to the available results of the uni-directional experimental tests of the XPS material [31].

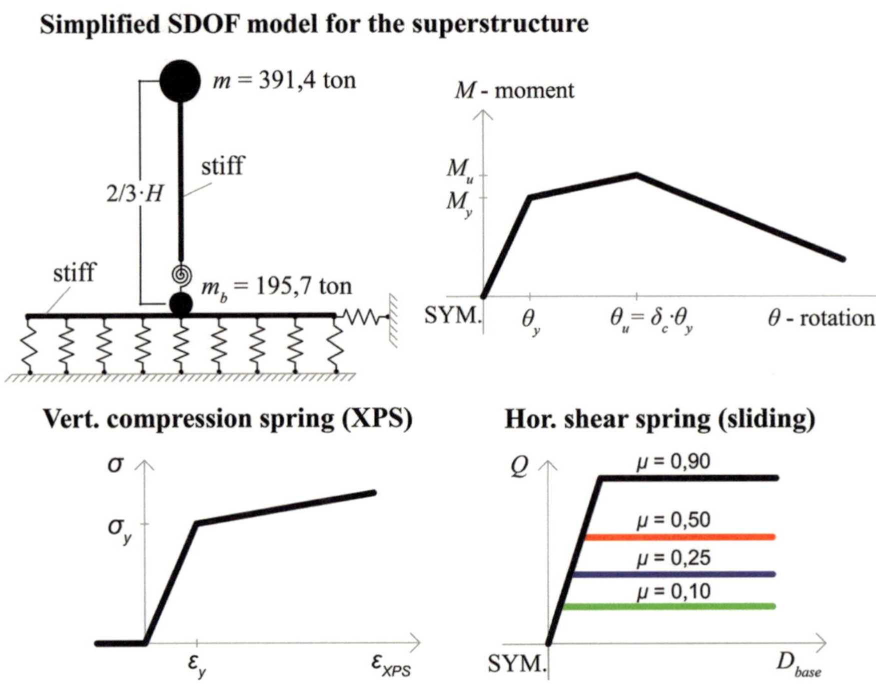

Figure 11. The numerical model considered in the parametric study.

In **Figure 12** fragility curves for the occurrence of sliding between the XPS boards are shown with the aim of illustrating the proposed seismic response scenarios. Sliding between the layers of XPS was numerically detected when the model reached the maximum resistance force of the horizontal spring (**Figure 11**). For each step of the IDA, the number of GMRs which caused sliding was calculated. The numerical probability of base sliding was then calculated as the ratio between the number of GMRs which caused sliding and the total number of analysed GMRs (30). The calculated numerical probabilities are shown in **Figure 12** in the form of stepped curves. Lognormal cumulative distribution functions were also fitted to the numerical probabilities. The fitted fragility curves were defined according to [51] and can be used to predict the seismic response scenario and to design the foundation detail accordingly.

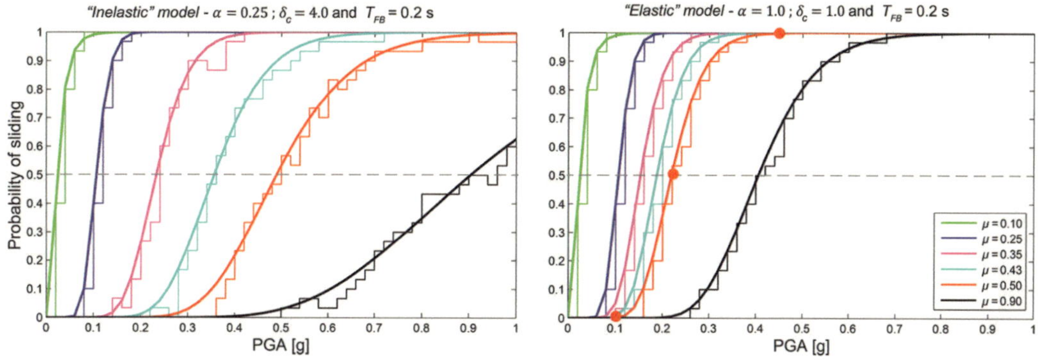

Figure 12. Fragility curves for the occurrence of sliding between the XPS boards.

Based on the presented fragility curves, an appropriate friction coefficient for the proposed foundation detail can be chosen in relation to the selected seismic response scenario. If the first seismic response scenario is selected, then the coefficients with zero probability of sliding will be the most appropriate solution for the detail. As can be seen from **Figure 12**, higher friction coefficients show zero probability of sliding for lower PGAs. On the other hand, models with lower friction coefficient values are vulnerable to sliding already at low PGAs. The third seismic response scenario, which is described as a sliding isolation system, could, on the other hand, be applied in the case of models with a 100% probability of sliding at the design PGA, whereas, for the second response scenario of controllable sliding, the use of a 50% probability of sliding at the design PGA is proposed. However, the probability of sliding for the second seismic response scenario could be selected individually by the building designer. If the desired level of protection of the superstructure is higher, the chosen probability of sliding will be closer to 100% and more similar to the third seismic response scenario. On the other hand, if the desired response at the design PGA is more similar to sliding prevention, the probability of sliding will be between 0 and 50%.

In the selected case of elastic models with μ = 0.50 the proposed seismic scenarios can be described as follows; (1) the first seismic response scenario would be allowed for PGA of less than approximately 0.10 g, (2) the second response scenario would be allowed for PGAs amounting to approximately 0.25 g and (3) the third response scenario would be required for PGAs greater than 0.45 g. It can be concluded from the above that different friction coefficients could be used for the sliding surface of the proposed foundation detail according to the selected seismic response scenario and the design PGA.

4.2.2. MDOF superstructure model

In order to verify the effectiveness of the proposed technological system in more detail, in the second stage of the study several variants of realistic multi-storey buildings were analysed. A typical passive or energy-efficient multi-storeyed RC office-building founded on a 30 cm thick RC foundation slab with a 2-layered XPS (the total thickness equal to 24 cm) beneath was used as a test example. The same building was also analysed in [52], where an evaluation of the critical parameters which could affect the seismic behaviour of buildings founded on an XPS layer can be found. In the same reference it is also possible to access all the other modelling input data, which are not reported there. In this chapter only selected results of the characteristic (2D) frame of the analysed building are presented (**Figure 13**). In the case of the selected concrete rectangular cross-sections of the beams and columns, the minimum amount of steel reinforcement for the selected ductility class medium (DCM) according to EC8 was adopted. Soil-structure interaction (SSI) effects were taken into account by assuming that the analysed multi-storeyed frames are founded on real soils (type A according to EC8). Since, in practice, no tensile resistance is provided by the soil-structure contact, the behaviour of the soil as well as the XPS in compression was modelled by nonlinear contact springs. The behaviour of the soil in shear was modelled by means of linear elastic springs, whereas the cyclic shear behaviour of the XPS was modelled by means of a kinematic hysteresis loop with a backbone, which is presented in (**Figure 11**), taking into account a sliding gap displacement (ΔH) equal to 5 cm, and assuming that the stiffness of the third branch was equal to the initial stiffness. Seismic analyses of the investigated frame systems were carried out by means of nonlinear dynamic response analyses, which were performed by the computer program SAP2000 [53]. The vertical loads which corresponded to the seismic limit state defined in EC8 were assumed as the initial loads in all the seismic analyses, in which a group of 7 real earthquake records was applied in one (horizontal) direction. These records were scaled to three different PGA levels (0.25 g, 0.375 g and 0.50 g). Details about these nonlinear dynamic analyses can be found in [52].

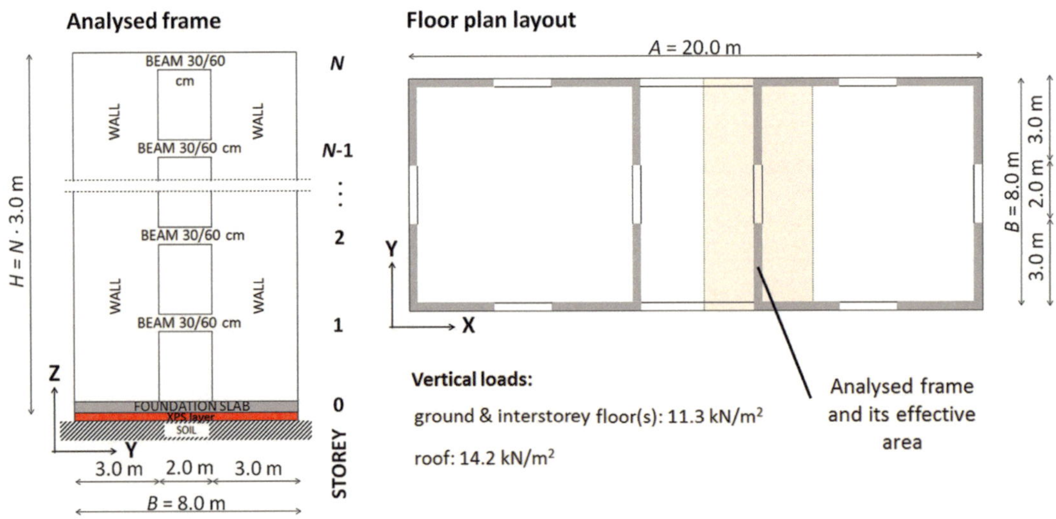

Figure 13. The investigated building's floor plan layout and the analysed 2D frame.

The seismic response of the selected analysed models which was obtained in the case of moderate seismic excitation (with a PGA equal to 1.5-times the design PGA) is presented in **Figure 14**. The typically obtained damage patterns of the 2- and 4-storeyed frames founded on foundation sets with different friction coefficients (μ) are presented, together with the absolute maximum base (D_{base}) and top displacements (D_{top}) as well as the maximum compressive deformations of the XPS layer (ε_{XPS}). The damage patterns show the performance levels of the superstructure (measured with reference to the calculated rotations in the generated plastic hinges).

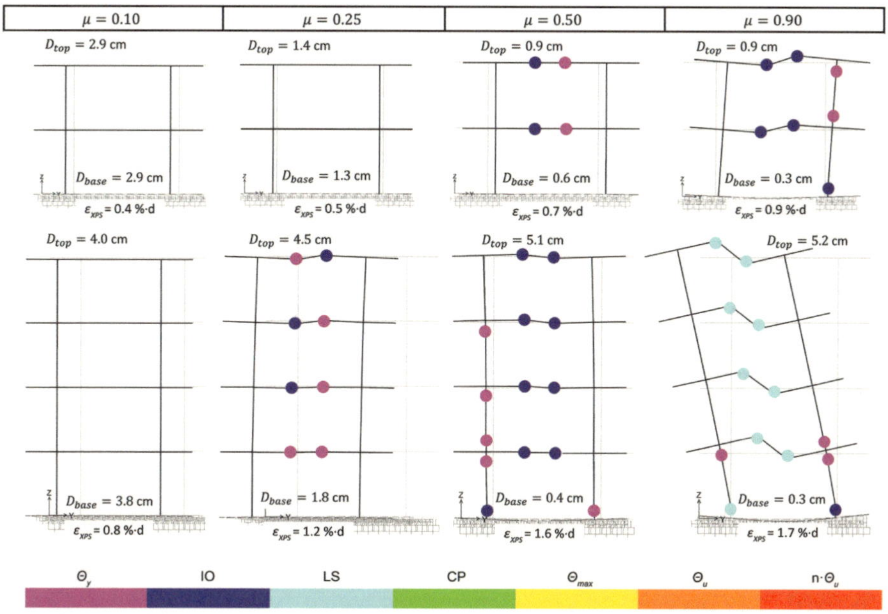

**The deformed shapes of the structures are presented (not to scale) for the last analysis step of a typical GMR.*

Figure 14. Typical damage patterns and average maximum displacements of the analysed two- and four-storeyed models subjected to PGA = 0.375 g.

It should be noted that, in all cases, the obtained base displacements (D_{base}) were smaller than the assumed gap width ($\Delta H = 5$ cm). It can be seen that, in the case of the analysed PGA level (0.375 g), a friction coefficient of $\mu \approx 0.4$ (for 2-storeyed model) and $\mu \approx 0.2$ (for 4-storeyed model) could be used if the third seismic response scenario (i.e., a sliding isolation system) were to be selected. In the case of higher values of the friction coefficient, the second (i.e., sliding controllable) or first (i.e., sliding prevention) seismic response scenario could be expected. In this connection, the limited damage state (i.e., with plastic hinges generated in the beams only) can be interpreted as acceptable for the second seismic scenario. In the analysed case the second scenario was reached when a friction coefficient of up to approximately $\mu = 0.5$ (for the 2-storeyed model), or approx. $\mu = 0.25$ (for the 4-storeyed model), was selected, whereas the first scenario occurred in the case of the higher friction coefficient values ($\mu \approx 0.9$ and $\mu \approx 0.35$ for the 2- and 4-storeyed models, respectively). It should be mentioned that, in the case of 4-storeyed model subjected to the design PGA level (0.25 g), the first (second) scenario occurred when a friction coefficient equal to approximately $\mu = 0.5$ (0.3) was selected, whereas the 2-storeyed models remained within the elastic range at the design PGA 0.25 g for all the considered values of μ. Furthermore, it should be noted that the response of slender frames with large height-to-width ratios (e.g., the analysed 6-storeyed frame, which – for the sake of brevity – is not shown) is governed strongly by the rocking mode of oscillation, which is evident from the obtained maximum edge compressive deformations of the XPS (ε_{XPS}). As can be observed also from the results presented in **Figure 14**, the sliding isolation system successfully protects the superstructure against a rocking mechanism, since, in the case of the 4-storeyed frame, ε_{XPS} decreases from 1.7% (sliding prevention) to 0.8% (sliding isolation system).

5. Conclusions

To successfully thermally insulate a building's foundations, a TI material with sufficient TI characteristics, compressive strength, water resistance, minimal long-term creep and good durability has to be used. In this chapter, the TI materials which are most commonly used beneath foundations (i.e., XPS, EPS, cellular glass, polyurethane and mineral wool) are presented, together with their basic mechanical properties. It can be seen that boards made of mineral wool have very low strength, stiffness and water resistance, so they are unsuitable to be used for this purpose. Boards made of polyurethane can also be deemed to be unsuitable because the production of this material can have negative effects on the Earth's ozone layer. Thus, three materials remain as being suitable for TI layers under foundations: XPS, EPS and cellular glass foam. Among them, at present boards made of XPS foam are the most commonly used due to their high water resistance, relatively high strength and competitive price.

The mechanical characteristics of XPS foam boards, which need to be known for the seismic analysis of buildings founded on XPS boards, were determined by means of laboratory tests. The results showed their measured compressive strength is, in general, always greater than the declared value, which defines the nominal class of the XPS. On the other hand, the obtained values of the elastic moduli are, as a rule, slightly lower than the declared values, which can be found in the producers' catalogues. The compressive and shear behaviour of XPS under monotonic and cyclic loading conditions has shown to have a very stable response, which is true for all the investigated test specimens. In general, the capacity to absorb energy in compression as well in shear is higher in the case of XPS material with a higher declared compressive strength. The values of the measured shear characteristics, which up to now have not been provided by the producers, were as follows: 0.14 (0.22) MPa (strength) and 4.5 (7.5) MPa (modulus) for XPS 400-L (700-L).

Besides its behaviour under monotonic and cyclic compressive as well as shear loading conditions, the sliding behaviour of differently composed TI foundation sets at different pre-compression levels was also investigated and their friction capacity estimated. Based on the results of an extensive parametric study of the seismic response of buildings [48], it was found that sliding between the individual components of TI foundation sets is a likely failure mechanism in the case of low-rise, light-weight and slender buildings subjected to seismic loads. The TI foundation set consisting of two XPS boards and a waterproofing layer (HI) in between (having adhesive on one side only) showed the smallest frictional capacity (the corresponding coefficient of friction at a pre-compression level of 50 kPa was around 0.3). It can be concluded that, in the case of TI foundation sets consisting of two XPS boards without a HI or PE sheet between them, the coefficient of friction amounted to 0.6, whereas coefficient of friction around 0.5 was obtained in tests of the contact between the XPS and the concrete. It was shown that the quality of the XPS boards (400 L and 700 L) did not significantly affect the frictional capacity of the analysed TI foundation set.

The results obtained in the performed numerical study of typical passive buildings have shown that foundation details, which permit sliding between the layers of TI boards by using low friction contact surfaces, can significantly reduce or even eliminate damage to the superstructure and thus act as a seismic fuse. The proposed seismic response scenarios (referred to as "sliding prevention", "sliding controllable" and the "implementation of a sliding isolation system") were demonstrated by means of nonlinear dynamic analysis for selected RC passive house structures with two and four storeys. Based on the fragility curves obtained by IDA for the occurrence of sliding between the layers of XPS boards, the likely seismic response scenario can be estimated depending on the PGA and the available friction coefficient. It is shown that lower values of the friction coefficient between the layers of TI boards reduce the level of damage to the superstructure. However, in this scenario large base displacements can occur, which must be taken into account when designing the size of the gap clearance, to prevent unfavourable pounding effects.

Based on all of the results obtained, it seems that the best solution is the implementation of a "sliding isolation system scenario", which works in a similar way to base isolation systems, and can provide full protection of the superstructure. However, from the financial point of view, this scenario usually requires some additional direct and hidden costs, which are frequently difficult for future owners to accept. To achieve this scenario more elaborate low friction material needs to be used, as well as more refined flexible installation systems, which could influence the final price of the newly built energy-efficient building. In the authors' opinion, at present the most feasible solution seems to be the "sliding controllable scenario", in which only easily available materials are needed. Such materials are already required under the foundations of modern energy-efficient buildings in order to prevent thermal bridges (they are governed by modern guides for low energy consumption). Thus, in this case the additional costs for horizontal stoppers and vertical restrainers (if required) would be low, as well as the costs of waterproofing (foil) materials with suitable low friction coefficients, which are in any case needed for all energy-efficient buildings, where they act as a waterproofing layer or a hydraulic barrier. The sliding controllable scenario can therefore be achieved without any significant additional financial costs and is therefore appropriate for use, on a wide scale, in the design of new energy-efficient buildings in earthquake-prone areas.

Acknowledgements

The financial support of the Slovenian Research Agency (Program No. P5-0068) is hereby gratefully acknowledged.

Author details

David Koren, Vojko Kilar and Boris Azinović*

*Address all correspondence to: boris.azinovic@fa.uni-lj.si

University of Ljubljana, Faculty of Architecture, Ljubljana, Slovenia

References

[1] European parliament and council. Directive 2010/31/EU of the European parliament on the energy performance of buildings. Brussels, 2010.

[2] Feist W. Wärmebrücken und Tragwerksplanung—die Grenzen des Wärmebrückenfreien Konstruirens. Darmstadt: Passivhaus Institut; 2007. 147 p.

[3] Dequaire X. Passivhaus as a low-energy building standard: contribution to a typology. Energy Efficiency. 2012;5(3):377–391. DOI: 10.1007/s12053-011-9140-8

[4] Praznik M, Butala V, Zbašnik Senegačnik M. Simplified evaluation method for energy efficiency in single-family houses using key quality parameters. Energy and Buildings. 2013;67:489–499. DOI: 10.1016/j.enbuild.2013.08.045

[5] Proietti S, Sdringola P, Desideri U, Zepparelli F, Masciarelli F, Castellani F. Life Cycle Assessment of a passive house in a seismic temperate zone. Energy and Buildings. 2013;64:463–472. DOI: 10.1016/j.enbuild.2013.05.013

[6] FIBRAN. Extruded polystyrene thermal insulation FIBRANxps [Internet]. 2015. Available from: http://www.fibran.com/files4users/files/documentation%20XPS/100_%5BEN_%5DLowRes.pdf [Accessed: 2015-12-5]

[7] Ramsteiner F, Fell N, Forster S. Testing the deformation behaviour of polymer foams. Polymer Testing. 2001;20(6):661–670. DOI: 10.1016/S0142-9418(00)00090-8

[8] Papadopoulos AM. State of the art in thermal insulation materials and aims for future developments. Energy and Buildings. 2005;37(1):77–86. DOI: 10.1016/j.enbuild.2004.05.006

[9] Bunge F, Merkel H. Development, testing and application of extruded polystyrene foam (XPS) insulation with improved thermal properties. Bauphysik. 2011;33(1):67–72. DOI: 10.1002/bapi.201110008

[10] Gnip IY, Vaitkus S, Keršulis V, Vėjelis S. Analytical description of the creep of expanded polystyrene (EPS) under long-term compressive loading. Polymer Testing. 2011;30(5):493–500. DOI: 10.1016/j.polymertesting.2011.03.012

[11] ROCKWOOL. Thermal insulation slab Rockwool [Internet]. 2015. Available from: http://www.rockwool.co.uk/products/u/2014.construction/11318/floors/ground-floors [Accessed: 2015-12-3]

[12] FOAMGLAS. Technische Daten FOAMGLAS Platten [Internet]. 2015. Available from: http://de.foamglas.com/de/waermedaemmung/produkte/produktuebersicht/foam-glas_platten/ [Accessed: 2015-12-12]

[13] JUBHome. Instructions for design of JUBhome base [Internet]. 2015. Available from: http://www.jub.si/jubhome-hise/nasveti/navodila-za-projektiranje-jubhome-base [Accessed: 2015-12-5]

[14] ELFOAM. Technical data ELFOAM boards [Internet]. 2015. Available from: http://www.elliottfoam.com/tech.html [Accessed: 2015-5-12]

[15] Diascorn N, Calas S, Sallée H, Achard P, Rigacci A. Polyurethane aerogels synthesis for thermal insulation—textural, thermal and mechanical properties. The Journal of Supercritical Fluids. 2015;106:76–84. DOI: 10.1016/j.supflu.2015.05.012

[16] Lyons A. Materials for architects and builders. London: Routledge; 2014. 512 p.

[17] CEN. European standard EN 826:2013: Thermal insulating products for building applications—Determination of compression behaviour. Brussels, 2013.

[18] CEN. European standard EN 1606:2013: Thermal insulating products for building applications—Determination of compressive creep. Brussels, 2013.

[19] ZRMK, AURE. Thermal insulating materials - thermal insulation of buildings [Internet]. 2003. Available from: http://www.aure.gov.si/eknjiznica/IL_2-03.PDF [Accessed: 2015-15-5]

[20] XPSA. Polystyrene based insulation boards products comparison [Internet]. 2015. Available from: http://www.xpsa.com/tech-info.html [Accessed: 2015-12-5]

[21] Méar F, Yot P, Viennois R, Ribes M. Mechanical behaviour and thermal and electrical properties of foam glass. Ceramics International. 2007;33(4):543–550. DOI: 10.1016/j.ceramint.2005.11.002

[22] Athanasopoulos GA, Pelekis PC, Xenaki VC. Dynamic properties of EPS geofoam: An experimental investigation. Geosynthetics International. 1999;6(3):171–194.

[23] Ossa A, Romo MP. Dynamic characterization of EPS geofoam. Geotextiles and Geomembranes. 2011;29(1):40–50. DOI: 10.1016/j.geotexmem.2010.06.007

[24] Trandafir AC, Bartlett SF, Lingwall BN. Behavior of EPS geofoam in stress-controlled cyclic uniaxial tests. Geotextiles and Geomembranes. 2010;28(6):514–524. DOI: 10.1016/j.geotexmem.2010.01.002

[25] Tomažič S, Logar V, Kristl Ž, Krainer A, Škrjanc I, Košir M. Indoor-environment simulator for control design purposes. Building and Environment. 2013;70:60–72. DOI: 10.1016/j.buildenv.2013.08.026

[26] François S, Schevenels M, Thyssen B, Borgions J, Degrande G. Design and efficiency of a composite vibration isolating screen in soil. Soil Dynamics and Earthquake Engineering. 2012;39:113–127. DOI: 10.1016/j.soildyn.2012.03.007

[27] Merkel H. Determination of long-term mechanical properties for thermal insulation under foundations. Buildings conference IX; USA, Atlanta: ASHRAE; 2004. p. 7.

[28] Vaitkus S, Vejelis S, Kairyte A. Analysis of extruded polystyrene short-term compression dependence on exposure time. Medziagotyra. 2013;19(4):471–474. DOI: 10.5755/j01.ms.19.4.2582

[29] Sadek E, Fouad N. Finite element modeling of compression behavior of extruded polystyrene foam using X-ray tomography. Journal of Cellular Plastics. 2013;49(2):161–191. DOI: 10.1177/0021955x13477436

[30] Abina A, Puc U, Jeglič A, Zidanšek A. Structural analysis of insulating polymer foams with terahertz spectroscopy and imaging. Polymer Testing. 2013;32(4):739–747. DOI: 10.1016/j.polymertesting.2013.03.004

[31] Kilar V, Koren D, Bokan-Bosiljkov V. Evaluation of the performance of extruded polystyrene boards—Implications for their application in earthquake engineering. Polymer Testing. 2014;40:234–244. DOI: 10.1016/j.polymertesting.2014.09.013

[32] Maleki S, Ahmadi F. Using expanded polystyrene as a seismic energy dissipation device. Journal of Vibration and Control. 2011;17(10):1481–1497. DOI: 10.1177/1077546309357693

[33] Gibson LJ. Biomechanics of cellular solids. Journal of Biomechanics. 2005;38(3):377–399. DOI: 10.1016/j.jbiomech.2004.09.027

[34] Chen W, Hao H, Hughes D, Shi Y, Cui J, Li Z-X. Static and dynamic mechanical properties of expanded polystyrene. Materials & Design. 2015;69:170–180. DOI: 10.1016/j.matdes.2014.12.024

[35] CEN. European standard EN 12090:2013: Thermal insulating products for building applications - Determination of shear behaviour. Brussels, 2013.

[36] Vejelis S, Gnip I, Vaitkus S, Keršulis V. Shear strength and modulus of elasticity of expanded polystyrene (EPS). Medziagotyra. 2008;14(3):230–233.

[37] Calderini C, Abbati SD, Cotič P, Kržan M, Bosiljkov V. In-plane shear tests on masonry panels with plaster: correlation of structural damage and damage on artistic assets. Bulletin of Earthquake Engineering. 2015;13(1):237–256. DOI: 10.1007/s10518-014-9632-y

[38] Azinović B, Kilar V, Koren D. Energy-efficient solution for the foundation of passive houses in earthquake-prone regions. Engineering Structures. 2016;112:133–145. DOI: 10.1016/j.engstruct.2016.01.015.

[39] Naeim F, Kelly JM. Design of seismic isolated structures: from theory to practice: John Wiley & Sons; 1999. 291 p.

[40] Hong WK, Kim HC. Performance of a multi-story structure with a resilient-friction base isolation system. Computers & Structures. 2004;82(27):2271–2283. DOI: 10.1016/j.compstruc.2004.06.002

[41] Panchal VR, Jangid RS. Seismic response of structures with variable friction pendulum system. Journal of Earthquake Engineering. 2009;13(2):193–216. DOI: 10.1080/13632460802597786

[42] Fadi F, Constantinou MC. Evaluation of simplified methods of analysis for structures with triple friction pendulum isolators. Earthquake Engineering & Structural Dynamics. 2010;39(1):5–22. DOI: 10.1002/eqe.930

[43] Koren D, Kilar V. The applicability of the N2 method to the estimation of torsional effects in asymmetric base-isolated buildings. Earthquake Engineering & Structural Dynamics. 2011;40(8):867–886. DOI: 10.1002/eqe.1064

[44] Becker TC, Mahin SA. Experimental and analytical study of the bi-directional behavior of the triple friction pendulum isolator. Earthquake Engineering & Structural Dynamics. 2012;41(3):355–373. DOI: 10.1002/eqe.1133

[45] Lu LY, Lee TY, Juang SY, Yeh SW. Polynomial friction pendulum isolators (PFPIs) for building floor isolation: An experimental and theoretical study. Engineering Structures. 2013;56:970–982. DOI: 10.1016/j.engstruct.2013.06.016

[46] Chung LL, Kao PS, Yang CY, Wu LY, Chen HM. Optimal frictional coefficient of structural isolation system. Journal of Vibration and Control. 2015;21(3):525–538. DOI: 10.1177/1077546313487938

[47] Polycarpou PC, Komodromos P. On poundings of a seismically isolated building with adjacent structures during strong earthquakes. Earthquake Engineering & Structural Dynamics. 2010;39(8):933–940. DOI: 10.1002/eqe.975

[48] Azinović B, Koren D, Kilar V. The seismic response of low-energy buildings founded on a thermal insulation layer – A parametric study. Engineering Structures. 2014;81:398–411. DOI: 10.1016/j.engstruct.2014.10.015

[49] Azinović B, Koren D, Kilar V. Principles of energy efficient construction and their influence on the seismic resistance of light-weight buildings. The Open Civil Engineering Journal. 2014;8:105–116. DOI: 10.2174/1874149501408010105

[50] CEN. Eurocode 8: Design of strucures for earthquake resistance—Part 1-1: General rules, seismic actions and rules for buildings. Brussels, 2005.

[51] Baker JW. Efficient analytical fragility function fitting using dynamic structural analysis. Earthquake Spectra. 2015;31(1):579–599. DOI: 10.1193/021113EQS025M

[52] Koren D, Kilar V. Seismic vulnerability of reinforced concrete building structures founded on an XPS layer. Accepted for publication in Earthquakes and Structures, An Int'l Journal. 2016.

[53] CSI. SAP2000 Structural and earthquake engineering software [Internet]. 2015. Available from: http://www.csiamerica.com/sap2000 [Accessed: 2015-12-3]

Chapter 5

Flammability of Bio-Based Rigid Polyurethane Foam as Sustainable Thermal Insulation Material

Mikelis Kirpluks, Ugis Cabulis and Andris Avots

Additional information is available at the end of the chapter

http://dx.doi.org/10.5772/62539

Abstract

One of the biggest disadvantages of rigid polyurethane foams is its low thermal resistance, high flammability, and high smoke production when burning. Greatest advantage of this thermal insulation material is its low thermal conductivity, which at 20–25 mW/(m·K) is superior to other commercially available insulation materials. In recent years polyurethane materials from renewable resources have been widely studied. But their use on industrial scale was limited due to inconstant performance and relatively high price of raw materials. Different bio-based raw materials, such as rapeseed oil and tall oil, could provide abundant feedstock for PU foam production. Decrease of flammability of PU materials conventionally is achieved by addition of flame retardants, halogen-containing compounds, and phosphates. It can be considered that halogenated fire retardants could have several health hazards, such as volatile compound emission from materials and toxic gas release during burning process. Expandable graphite could be an answer to this flammability problem. This chapter describes development of bio-based rigid polyurethane foams and their flammability reduction using sustainable flame retardants. Different expandable graphite intumescent flame retardants provided significant flammability reduction while maintaining low thermal conductivity of insulation materials.

Keywords: rigid polyurethane-polyisocyanurate foam thermal insulation, bio-based polyols, thermal insulation, flammability, intumescent flame retardants, expandable graphite

1. Introduction

1.1. Development of bio-based rigid polyurethane foams

Polyurethane (PU) materials have a broad range of applications such as flexible foams, rigid foams, thermoplastic elastomers, coatings, and adhesives. About 29% of the PU material industry is attributed to rigid PU and polyisocyanurate (PIR) foam production [1]. These materials are mainly used as heat-insulating materials in civil engineering and the refrigeration industry due to their low thermal conductivity (λ). The industry standard of λ for rigid PU/PIR foams is 0.020 ± 0.002 W/mK, which, compared to the case of other insulation materials such as extruded polystyrene (XPS), expanded polystyrene (EPS), glass, or mineral wool, is almost twice lower [2]. Low λ values of rigid PU/PIR foams allow their application of in fast-growing markets like the nearly zero energy building industry. An additional advantage over XPS/EPS foams is the sustainability of PU/PIR materials as they can be obtained from renewable resources.

Sustainable solutions have been studied across a whole range of polymer materials [3–6]. Emphasis has been placed on PU materials because at least one component of a polymer material can be obtained from renewable resources [3–13]. Unfortunately, the majority of PU materials are produced from petroleum-based feedstocks. In recent years, the prices of crude oil and petroleum-based feedstocks have fluctuated significantly. Although the current prices of crude oil are the lowest in the recent decade [14], it is important to find a suitable substitute to petrochemical raw materials as they will run out eventually. Also the use of renewable resources in polymer materials decreases the global warming potential of the material and can sequester a large amount of CO_2 from the atmosphere while improving the impact the material has on the environment [15]. This has led to an increasing need and interest in developing alternate polyols (e.g., BASF castor oil-based Balance™, Cargill soybean-based BiOH™, and Dow soybean-based Renuva™) based on raw materials from renewable resources that offer reduced prices with a more favorable environmental footprint compared to the current petroleum-derived polyols [16–18].

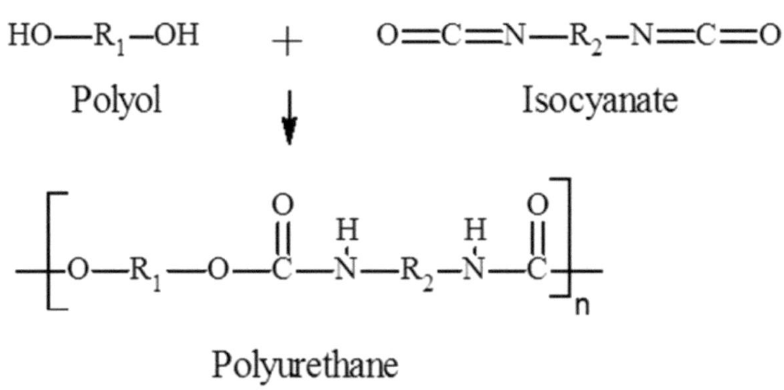

Figure 1. Generic scheme of PU material synthesis.

http://dx.doi.org/10.5772/62539

PU materials are obtained in the chemical reaction of hydroxyl derivatives—polyols and isocyanates as shown in **Figure 1**. For rigid PU foam synthesis, usually polyols with higher functionality (f_n = 2.5–5.0) and polymeric diphenylmethane diisocyanate (pMDI) are used.

Renewable materials are usually introduced into the PU polymer matrix as a polyol—a hydroxyl derivative of vegetable and other natural oils. Bio-based polyols have been investigated for some time, and studies in this field are currently continued. Good quality polyols have been obtained from different vegetable oils such as rapeseed oil (RO), castor oil, palm oil, and especially soybean oil [3–13]. Most of these oils are already used to produce a PU material feedstock on an industrial level [16–18] and can be potential replacements for petrochemical polyols in the synthesis of rigid PU foams.

Industry uses renewable raw materials not only because of the sustainability and marketing aspect of products. These materials provide a competitive and commercially viable alternative to petrochemical resources [19]. Vegetable oil polyols also provide additional positive properties of a PU material, such as higher hydrophobicity [11].

Vegetable oils are triglycerides of three different fatty acids that are joined together by the glycerol backbone as can be seen from **Figure 2**. Most of common natural oil triglycerides contain fatty acid chains that vary from 14 to 22 carbons in length, with 0–3 double bonds per fatty acid [3, 8, 10].

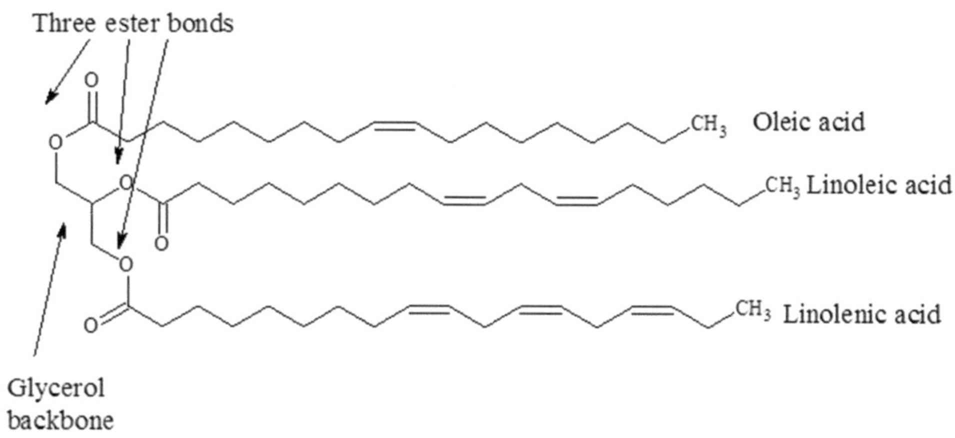

Figure 2. Generic structure of vegetable oil containing oleic, linoleic, and linolenic acid chains [11].

Before being applied in PUR systems, hydroxyl groups have to be introduced into structure of vegetable oil [6]. A number of methods to synthesize natural oil-based polyols are known, e.g., transesterification with polyfunctional alcohols, hydroformylation and hydrogenation, epoxidation followed by oxirane ring opening, microbial conversion, thermal polymerization followed by transesterification, ozonolysis and hydrogenation, or halogen addition and nucleophilic substitution [20, 21]. This chapter will describe polyols and rigid PU/PIR foams obtained from TO and RO using transesterification with polyfunctional alcohols [9, 10] and epoxidation followed by oxirane ring opening [10–13].

Unfortunately, most of the mentioned vegetable oils can be classified as first-generation bio-based raw materials [22, 23]. This means that the production of these polyols is competing with the production of food. TO can tackle this question because it is a by-product of cellulose production—a product of forest biomass processing and not an agricultural product. TO is a mixture of fatty and rosin acids, the generic structures of which can be seen from **Figure 3** [24–27]. TO, also like RO, should be chemically modified by introducing two or more hydroxyl groups in order to use it as a raw material for PU production.

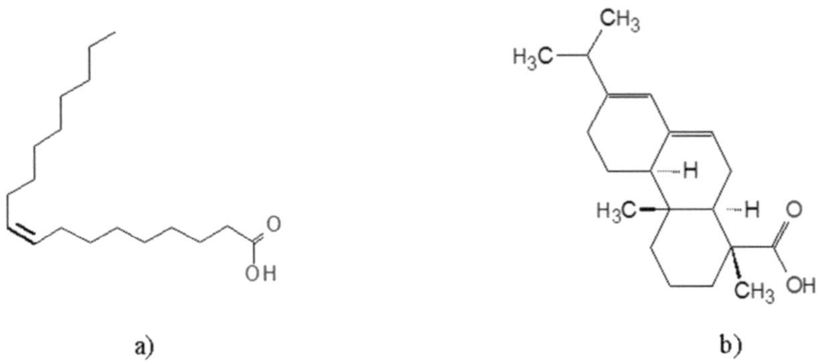

Figure 3. Basic structure of TO components: (a) fatty acid (oleic acid) and (b) rosin acid (abietic acid) [24, 25].

A renewable polyol from TO can be synthesized using a method developed at the Latvian State Institute of Wood Chemistry by the esterification of TO fatty and rosin acids with triethanolamine (TEOA). Distilled TO with a rosin acid content of 20% (Forchem, Finland) was used as a raw material for bio-based polyol development. **Figure 4** shows the TO esterification with TEOA [24, 25].

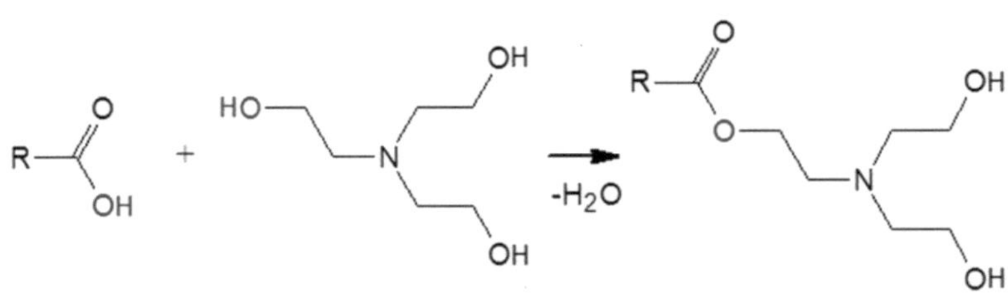

Figure 4. Reaction scheme of TO esterification with TEOA, where R is specific carboxylic acid radical [24, 25].

1.2. Flammability of rigid PU foams

One of the major disadvantages of rigid PU foams is their low thermal resistance, high flammability, and high smoke production when burning. PU foams based on petrochemical and bio-based polyols are ignitable and can be an additional fuel source in the case of a fire disaster. This is a serious concern and restricts the PU material application [28]. The limiting oxygen index is the minimum concentration of oxygen in material surroundings that will support combustion of a polymer. It is measured by passing a mixture of oxygen and nitrogen over a burning specimen and reducing the oxygen level until a critical level is reached. The limiting oxygen index of non-modified PU foams is in the range of 16–18 [29]. Highly porous lightweight combustible foams tend to have a fast flame spread and a high heat emission. The increasing demand for PU foams is the reason why many studies are devoted to fire retardancy [30, 31].

An improvement in the thermal stability of PU foams may be achieved through the introduction of isocyanurate trimerization structures into the PU matrix [11], the so-called PIR foams. The trimerization reaction of isocyanate groups is shown in **Figure 5**. Isocyanurates, from a thermodynamic point of view, are more thermally stable than urethane bonds (urethane dissociates at approx. 200°C as opposed to 350°C for PIRs). The thermal stability of isocyanate-based polymers is provided in the following order: isocyanurate (350°C) > urea (250°C) > urethane (200°C) > biuret (135–140°C) > allophanate (106°C) [32, 33].

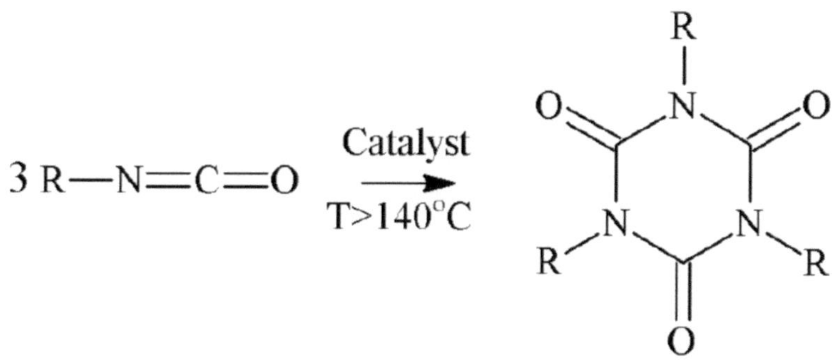

Figure 5. Trimerization reaction of three isocyanate groups.

PIR foams have higher fire retardancy but their broader use is limited due to price and technological difficulties. PIR foams are obtained in large excess of isocyanate (molar ratio of OH and isocyanate groups: 1/1.8–1/6), which means more complicated processing equipment as the volume ratios of components are not equal. Also the isocyanate trimerization reaction occurs only when the temperature of the reaction mixture is above 140°C. Thus, thermal insulation production via spray foaming in situ is difficult and rarely used commercially [33].

The decrease of flammability of PU/PIR materials is conventionally achieved by the addition of different flame retardants (FRs)—compounds, containing halogen, phosphorus, nitrogen, etc. Moreover, halogenated FR is the subject of considerable discussion in the industry and scientific community. It can be considered that halogenated FR could have several health hazards, such as volatile compound emission from materials and toxic and hazardous gas release during the burning process [34, 35].

Intumescent halogen-free FRs can be good substitutes for halogenated ones due to the formation of a protecting char layer that covers the surface of a material and limits the amount of heat reaching the polymer, slowing down the thermal degradation of the material. Also the protective char layer decreases the transition of volatile compounds into the gas phase, thus reducing the amount of fuel in the gas phase and decreasing the emitted heat of the fire [36, 37]. Moreover, such an intumescent layer makes a thermal barrier, protecting the foam core from high temperature, and is used as a smoke suppressant [38, 39]. Expandable graphite (EG) can be used as a good substitute for halogenated liquid FR in rigid PU/PIR foams [40].

EG has a special graphite flake structure, i.e., stacked layers of hexagonal sp^2-hybridized carbon structures. Intercalated EG with H_2SO_4 is shown in **Figure 6**. EG material can be treated with sulfuric acid, nitric acid, or acetic acid, which are intercalated into the graphite crystalline structure between carbon flakes. When heat is applied, the acid releases gas that expands or exfoliates the graphite particles.

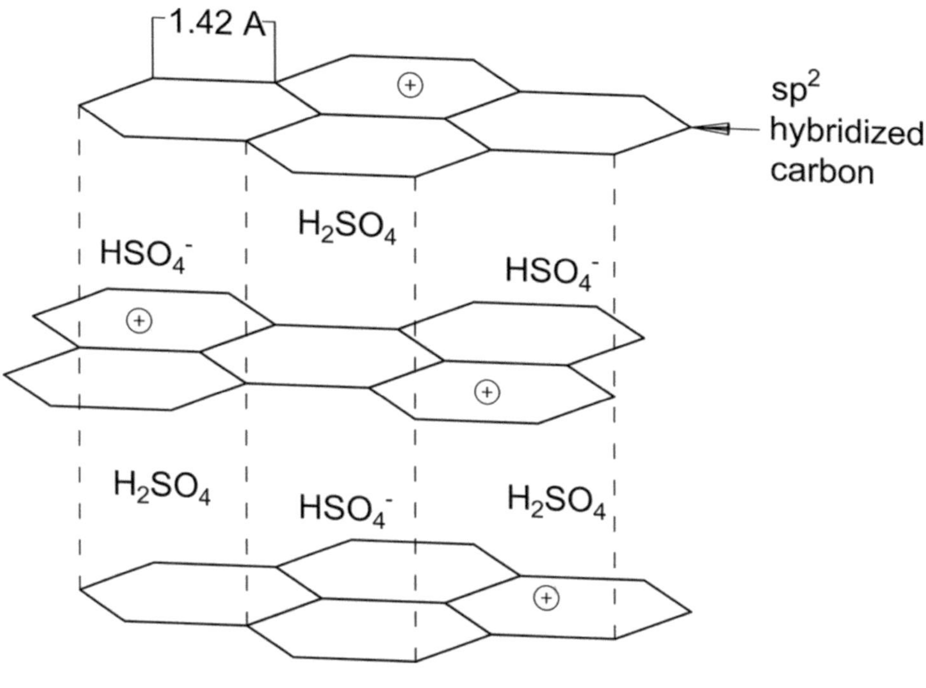

Figure 6. Generic structure of EG intercalated with H_2SO_4 [40].

High temperatures cause the oxidization of graphite in the following reaction with H_2SO_4:

$$C + 2H_2SO_4 \rightarrow CO_2\uparrow + 2H_2O\uparrow + 2SO_2\uparrow$$

The released gases cause the expansion of graphite, which acts as a physical barrier for the heat and mass transfer from and into polymer material [41]. EG forms a stable char insulating layer structure on the surface of the material that prevents heat and volatile compound transfer. Volatile compound reduction in gas phase means less fuel for burning process. Also, limiting the heat transfer to material means longer degradation of polymer matrix, which in result limits the volatile compound release. Ultimately, the carbon char layer prevents the PU/PIR foams from burning [42, 43]. Additional benefit of such FR is reduction of smoke as the protective char layer also stops particle escape into airflow.

EG keeps most of natural graphite attributes, such as low price and high porosity, making it very useful as functional carbon material that can be applied in various fields of the polymer industry, but especially as FR [44]. EG addition with loads over 20% into rigid PU foams with an Apparent bulk density of 35 kg/m^3 gives a significant improvement in the fire resistance of the materials [45]. The FR efficiency was enhanced when the apparent density of PU/PIR foams was increased at a fixed EG loads. Also with an increase in the EG content at a fixed apparent PU/PIR foam bulk density, the flammability was reduced [35, 39]. The size of EG particles also influences the FR efficiency. EG particles of a smaller size did not produce enough char to cover the whole surface of the burning sample, resulting in a poor fire-retardant property of rigid PU/PIR foams [46].

Although EG can replace halogenated FR and significantly decrease the flammability of polymer materials, there is a significant drawback regarding thermal insulation materials. Graphite is an excellent heat conductor, so with the addition of EG into PU/PIR foams, the thermal conductivity of the material rises, which is a highly undesired property for thermal insulation material [47]. To avoid the abovementioned disadvantages, IF (nonwoven glass fiber fabric filled with EG) can be used as a FR solution. Recently, several researchers have investigated the fire behavior of the foam/fabric combination [48, 49] using a cone calorimeter, but it was done only for flexible PU foams. The idea was to test the fire behavior of mattresses for automotive and furniture cushions; the results have shown that such fabric is a viable fire-retardant solution.

The goal of this study was to develop a thermal insulation material from renewable resources with competitive thermal insulation and fire retardancy properties and replace halogenated FR with more sustainable solutions. For this purpose, several bio-based polyols were used. TO, a by-product of cellulose production, was used to synthesize the raw material needed for PU/PIR foam development. TO was chosen as a cheap and available raw material, which is not competing with food production as other vegetable oils that are agricultural products. The bio-based polyol from TO was synthesized using the esterification reaction with polyfunctional amine-based alcohol—TEOA. RO was used as the second renewable material feedstock. Polyols for rigid PU/PIR foam production were obtained by two synthesis methods. The RO polyol was synthesized similarly to the TO polyol by transesterification of the triglyceride structure of RO with TEOA. The third type of the bio-based polyol was also obtained from RO, but double bonds in the RO chemical structure were targeted to introduce hydroxyl groups into the compound. The cumulative effect of the introduction of isocyanurate structures into the PU matrix and EG as additive FR is a prospective solution to decrease the flammability of rigid PU/PIR foams. The second approach to reduce the flammability of PU/PIR foams—protection of insulation materials with (IF), to keep excellent rigid PU/PIR foam insulation properties—was also investigated. Two different additive liquid FRs—tris(2-chloropropyl) phosphate (TCPP) and dimethyl-propyl-phosphate (DMPP)—were compared for their influence in the decrease of PU/PIR foam flammability.

2. Development of bio-based rigid PU foams

2.1. Bio-based polyols

The most significant properties of the polyols used in this study are presented in **Table 1**. TO polyol and RO polyols were synthesized at Institute of Wood Chemistry [10, 25, 26]. The epoxidized RO polyol was synthesized at the Cracow Technical University [11–13].

Polyol type	OH value, mgKOH/g	Viscosity at 25°C, mPa·s	Acid value, mgKOH/g	M_n, g/mol	f_n	Water content, %
TO polyol	342	280 ± 25	<5	391	2.39	0.24
RO polyol	301	190 ± 25	<5	474	2.55	0.05
RO Epoxy	276	2260 ± 50	<5	635	3.12	0.36

Table 1. Technical data of renewable polyols used for PU/PIR foam synthesis.

Gel permeation chromatography was used to determine the values of the number-average molecular weight (M_n) and number-average functionality (f_n) of the obtained polyols. The number-average functionalities of polyols were calculated based on hydroxyl values, and M_n was experimentally determined (Eq. (1)) [10]:

$$f_n = \frac{M_n \cdot OH_{val}}{56110} \quad (1)$$

where f_n is the number-average functionality, M_n is the number-average molecular weight, and OH_{val} is the hydroxyl value of polyol.

Viscosity of polyols was determined according to the DIN 53015 standard at 25°C using a falling ball viscometer KF 100 (RheoTec Messtechnik GmbH, Germany). Water content of polyols was determined according to the DIN 51777 standard.

The chemical structure of the used polyols was studied by Fourier transform infrared spectroscopy (FTIR) measurements. **Figure 7** presents FTIR spectra of polyols from renewable materials. Peaks at 3444–3385 cm^{-1} indicated the presence of OH groups in the polyols chosen for this project. The TO polyol and the RO polyol showed double-bond stretching at 3008 cm^{-1}. In contrast, the RO Epoxy polyol did not have this peak as double bond was transformed in the epoxidation and ring-opening reaction. TO and RO polyol FTIR spectra were quite similar as expected, also because of the similarity of the chemical structure. Tertiary amine group vibrations are seen at 1043–1042 cm^{-1}. The RO Epoxy polyol differs from the TO and RO polyols with ether bond —C—O—C— symmetric stretching at 1103 cm^{-1}, which is present due to epoxy ring opening with diethylene glycol. The lack of epoxy ring vibration at the 928 cm^{-1} peak confirms the ring-opening reaction. The addition of tertiary amine groups into the polyol could make it more catalytically active as most commercial PU catalysts are amine based. This could be a beneficial property because catalysts are among the most expensive components in PU foam formulation. The TO polyol should contain aromatic structures because distilled TO with 20% of rosin acids was used to synthesize the TO polyol but no peaks associated with aromatic rings were noticed as the concentration of aromatic groups in the end product was too low. All natural oil polyols showed a C═O bond stretching peak at 1760–1665 cm^{-1} [10, 50, 51].

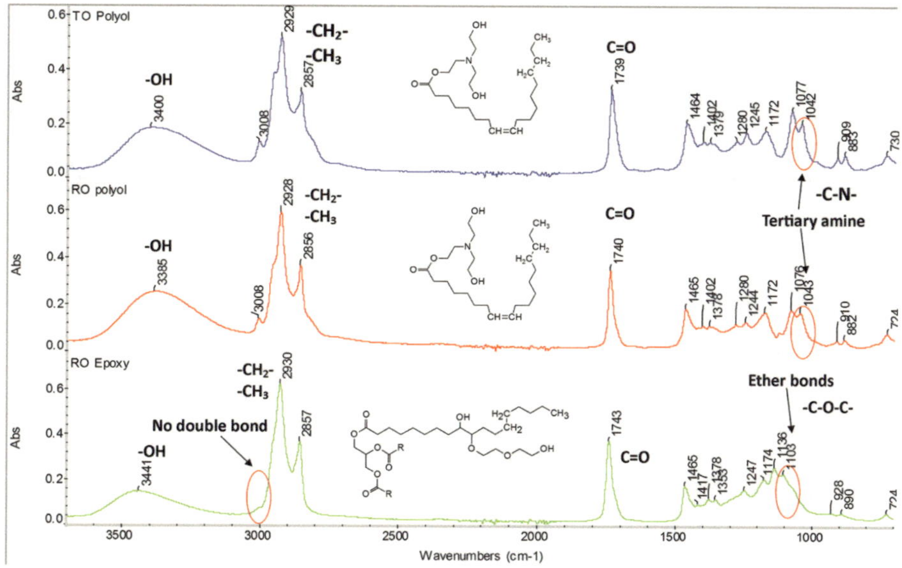

Figure 7. FTIR spectra for polyols from renewable raw materials.

2.2. Rigid PU/PIR foam sample preparation and materials used

2.2.1. Materials

Polyols from renewable resources were used as the base of the rigid PU/PIR foam polyol component. A higher functional polyether polyol based on sorbitol Lupranol 3422 (contains only secondary hydroxyl groups, OH value 490 mg KOH/g) from BASF was added to increase the cross-linkage density of the polymer matrix. An additive surfactant NIAX Silicone L6915 was used to obtain closed-cell PU/PIR foams. The reactive delayed action time amine-based catalyst NP-10 available from Momentive Performance Materials Inc was used. TCPP and DMPP from Lanxess Deutschland GmbH were used as additive liquid FRs. Distilled water was used as a chemical blowing agent. Polymeric diphenylmethane diisocyanate—IsoPMDI 92140 (pMDI) from BASF—was used as an isocyanate component (NCO = 31.5 wt%). EG with the trade mark EG 290 from Sinograf, Poland, was used as intumescent FR. Carbon content in EG was 90%, particle dimensions 0.2–0.6 mm, expansion of EG 200–400 ml/g, moisture content ~1.5%, and bulk density ~0.65 g/m^3.

IF from Technical Fibre Products Ltd. was used for the development of sandwich-type PU foam panels: flexible, thermally expandable IF (60152C) which is based on mineral and glass fibers with incorporated graphite particles. The activation temperature of this fabric is >190°C, at which it starts to expand and reaches expansion ratio of 9:1 at 450°C. The thickness of non-expanded fabric is 2.00 mm; typical density 220 kg/m^3 and tensile strength 175 N/15 mm.

2.2.2. *Rigid PU/PIR foam sample preparation*

The polyol component was obtained by weighting all necessary components (polyol, cross-linkage reagent catalyst, blowing agent, surfactant, 7% of liquid FR, and different amounts of EG) and stirred for 1 min by a mechanical stirrer at 2000 rpm. Then, an appropriate amount of pMDI was added to obtain PU/PIR foams with respective isocyanate indices (*II*=110, 150, 200, and 250) and both components were stirred. Isocyanate index is the ratio of the equivalent amount of isocyanate used relative to the theoretical equivalent amount times 100. The reacting mass of PU foam was poured into a stainless steel closed-type mold, preheated to 50°C. For final curing of PU foams, the mold was placed into an oven at 50°C for 2 h. PU/PIR foams with IF were obtained by placing the fabric on the bottom of the stainless steel mold and pouring the PU/PIR reacting mass over it. In this way one side of the PU/PIR foam material became protected; this side was also tested in the reaction to heat flux tests. Four types of the obtained PU foam samples are unmodified PU/PIR foam, PU/PIR foam filled with different amounts of EG (3%, 6%, 9%, and 15%), PU/PIR foam protected with IF and PU/PIR foam with EG (3%, 9%, and 15%), and IFs that are shown in **Figure 8**.

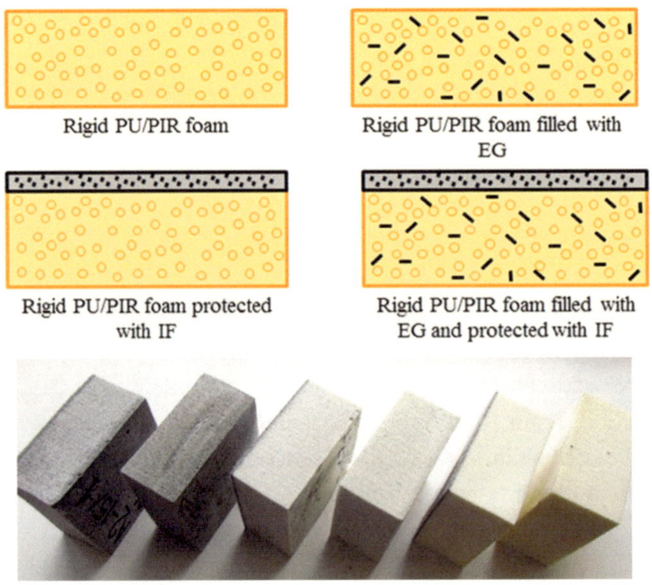

Figure 8. Four types of PU/PIR foam samples studied [47].

Samples for different tests from the prepared PU/PIR foam blocks were cut using a band saw and then conditioned for at least 24 h at room temperature. The apparent bulk density of the obtained PU/PIR foams was tested according to the ISO 845 standard. The reaction to the 35 kW/m^2 heat flux was tested using a FTT Dual Cone Calorimeter from Fire Testing Technology Ltd. The peak heat release rate (pHRR, kW/m^2), time to pHRR (TTP, s), ignition time (IT, s), time to flame out (FOT, s), total smoke release (TSR, m^2/m^2), and maximum average rate of heat emission (MARHE, kW/m^2) were determined by this equipment according to the ISO 5660 standard. The distance from the heater to the sample surface was set at 25 mm and the horizontal sample holder was used. The thermal conductivity coefficient (λ, mW/m·K) was tested using a Linseis Heat Flow Meter according to the ISO 8301 standard. Compression strength and modulus of elasticity for PU foams parallel and perpendicular to the foaming direction were tested according to the ISO 844 standard with one offset of the sample size; cylinders with a diameter of 20 mm and a height of 22 mm were tested. Mechanical testing of PU/PIR foam was performed on testing machines Zwick Roell 1000 N. Scanning electron microscopy (SEM) was applied to obtain images of PU/PIR foams and to determine the material morphology. SEM microscope, Tescan TS 5634, was used. The thermal stability was tested via thermogravimetric analysis (TGA) using a Mettler-Toledo TGA/SDTA 851e under a nitrogen flow and at a heating rate of 10°C/min from room temperature to 1000°C.

3. Properties of bio-based rigid PU foams

3.1. Thermal conductivity of the developed rigid PU/PIR foams

The main characteristics of any thermal insulation material are thermal conductivity and thermal resistance. **Figure 9** shows the measured λ for the rigid PU foam obtained from the TO polyol and filled with different amounts of EG particles. The isocyanate index of these samples was 110. **Figure 9(a)** is for the PU foam without DMPP liquid FR and the image (b) is for the PU foam with the addition of 7% of DMPP. λ values of samples with IF protection were also measured. It can be seen that IF has no effect on the λ of PU foams, but there is a clear increase of thermal conductivity with the addition of EG. The λ of the PU foam with no DMPP FR is 24.44 mW/m·K, which is decent for the material that could be used in building construction. Unfortunately, the addition of DMPP increased λ up to 28.74 mW/m·K which is only marginally satisfactory, because at this value the thermal insulation properties of the developed PU foam are in the range of XPS/EPS. Rigid PU foams cannot compete with XPS/EPS in terms of price even if they are produced from renewable resources. The increase of the developed PU foam thermal conductivity can be explained by the plasticization of the PU polymer matrix. Most of conventional additive liquid FRs decrease the glass transition temperature of the PU polymer matrix [52]. They swell the network structure of the PU matrix and dilute aromaticity. Due to the decrease of the cross-linkage density, blowing agent can defuse more easily through the PU foam cells even though the developed PU foam has a closed-cell structure. The closed-cell content of all developed PU foams was >90%. The substitution of CO_2 or another gas (blowing agent) with air in conventional PU foams is a slow process and takes years [53]. However, even with λ 28.74 mW/m·K, the developed PU foams are an energy-efficient thermal insulation material.

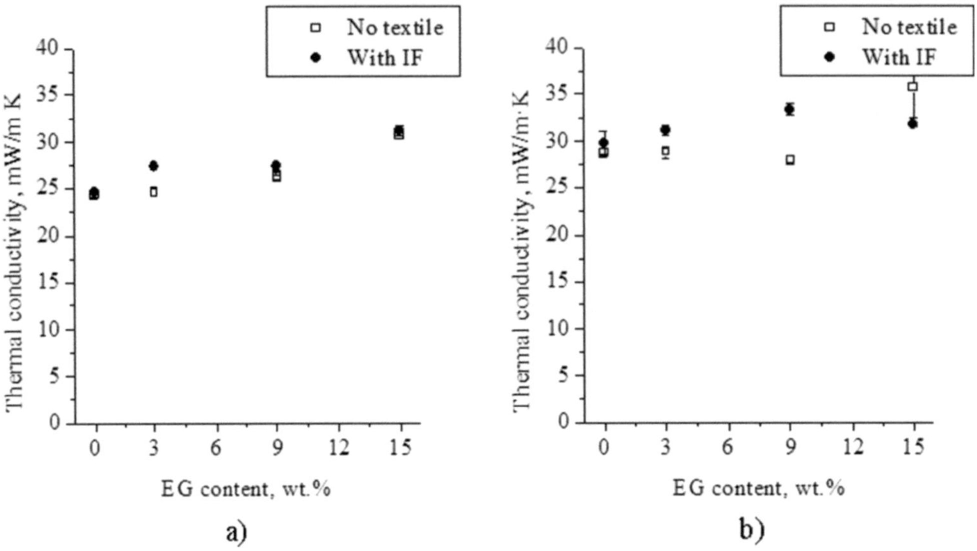

Figure 9. Thermal conductivity of rigid PU foams (II-110) protected with IF: (a) with no liquid FR but with EG and (b) with DMPP as FR and EG.

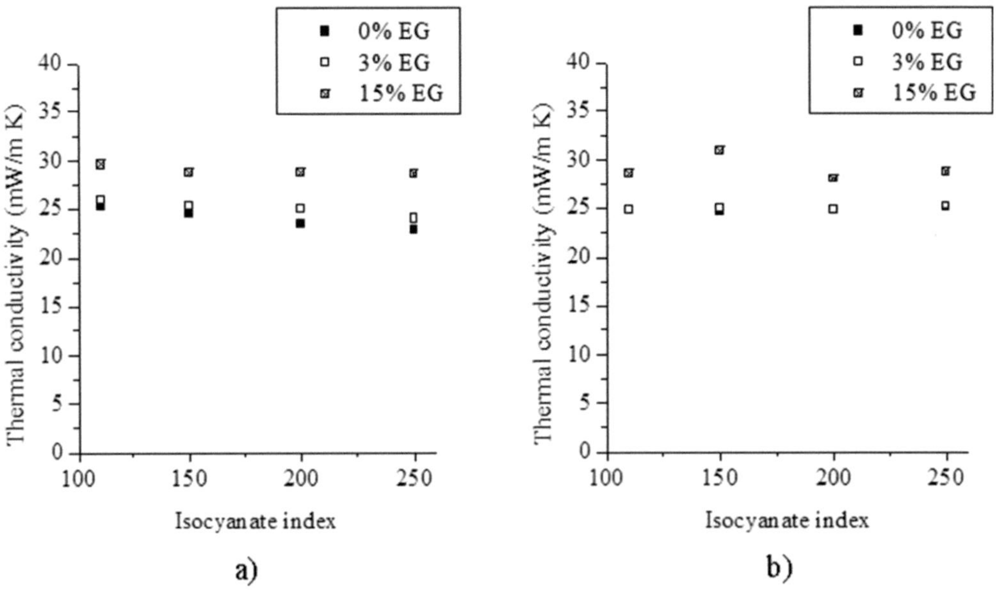

Figure 10. Thermal conductivity of rigid PU/PIR foams with different isocyanate indices and different EG content: (a) with no IF and (b) protected with different IF.

http://dx.doi.org/10.5772/62539

Figure 10 shows λ values of rigid PU/PIR foam with different isocyanate indices and different EG content. Also in the case of different isocyanate indices, IF had little influence on the thermal conductivity of the developed materials, and λ values increased with increasing EG content. The lowest value of λ was 22.80 mW/m·K in the case of rigid PIR foam with the highest isocyanate index (250). Due to cyclotrimerization, shown in **Figure 5**, the cross-linkage density of the polymer matrix and aromaticity increased. This helps reduce the CO_2 diffusion through cell walls. The λ value of 22.80 mW/m·K can be considered up to industrial quality standards of a rigid PU/PIR foam thermal insulation material.

Also for rigid PU/PIR foams obtained from RO bio-based polyols, the EG increased the λ value. Similarly to the TO polyol, the PU/PIR foam with a higher isocyanate index had a lower λ value, and the PU/PIR foam based on the RO Epoxy polyol showed the best insulation properties with l value of 22.79 mW/(m×K) (**Figure 11**).

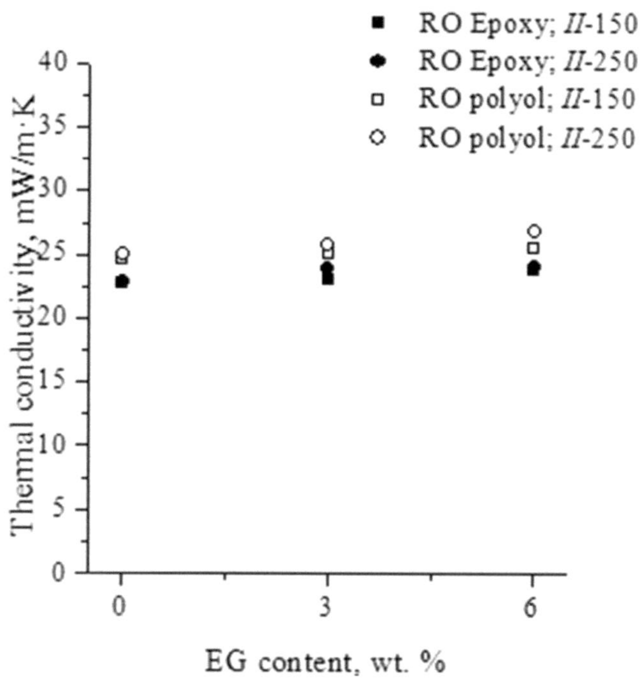

Figure 11. EG influence on the thermal conductivity of rigid PU/PIR foams based on different RO polyols.

It can be seen that EG increases the λ; at 15% of EG in the PU foam with liquid FR, λ increases from 28.74 to 35.72 mW/m·K. The carbon structure of EG is a good heat conductor, hence the increase of λ. The best-case scenario could be if it would be possible to obtain a PU foam composite with low λ and high flame resistance properties. Further analysis of the obtained PU/PIR foams will show that it is possible to find a compromise between these characteristics using IF and EG as FR systems.

3.2. Mechanical and morphological properties of the developed rigid PU/PIR foams

Other characteristics of the developed TO polyol-based PU foams with different amounts of EG are listed in **Table 2**. It can be seen that the addition of EG increased the apparent bulk density of the developed PU foams, but this did not improve the mechanical properties of PU foams. PU foam panels with the apparent bulk density ~ 60 kg/m^3 are used as thermal insulation in floor panels. The increase of apparent bulk density also correlates with the increase of λ. EG does not have a reinforcement effect. Even at 15% of EG in PU foam, the particles are too far spread out. SEM images show that, although the particles have good phase incorporation into the PU matrix, they disrupt the cell morphology. This is the reason for the decrease of mechanical properties.

Foam characteristic				
Rigid PU/PIR foams based on TO polyol with no liquid FR				
EG content, %	0	3	9	15
Apparent bulk density, kg/m^3	58.9 ± 1.5	59.9 ± 1.9	61.9 ± 3.3	65.5 ± 3.7
Compression strength Z, MPa	0.38 ± 0.04	0.36 ± 0.03	0.35 ± 0.02	0.29 ± 0.03
Young's modulus Z, MPa	9.4 ± 1.5	8.7 ± 1.6	8.7 ± 0.8	7.1 ± 1.2
Compression strength X, MPa	0.22 ± 0.03	0.24 ± 0.02	0.21 ± 0.02	0.22 ± 0.02
Young's modulus X, MPa	4.2 ± 1.0	4.4 ± 0.7	4.1 ± 0.5	4.6 ± 0.6
Rigid PU foams based on TO polyol with DMPP				
Apparent bulk density, kg/m^3	67.9 ± 2.3	70.0 ± 2.1	73.1 ± 3.4	68.6 ± 2.4
Compression strength Z, MPa	0.35 ± 0.03	0.45 ± 0.03	0.32 ± 0.03	0.32 ± 0.02
Young's modulus Z, MPa	7.4 ± 1.0	11.5 ± 1.6	7.0 ± 1.3	8.4 ± 0.6
Compression strength X, MPa	0.26 ± 0.01	0.35 ± 0.02	0.24 ± 0.01	0.17 ± 0.01
Young's modulus X, MPa	4.2 ± 0.3	6.8 ± 0.7	4.3 ± 0.3	2.5 ± 0.4

Z-measurements of compression strength parallel to foaming direction
X-measurements of compression strength perpendicular to foaming direction

Table 2. Bulk density, compression strength, and Young's modulus for PU foams with no liquid FR but with EG and for PU foams with DMPP as FR and EG.

The compression strength and Young's modulus of rigid PU/PIR foams with different isocyanate indices and different EG content can be seen from **Figure 12**. The increase of isocyanate index also improved the mechanical properties of the developed insulation materials. The highest values of compression strength and Young's modulus (0.49 and 10.34 MPa, respectively) were achieved for rigid PU/PIR foams with an isocyanate index of 200. The improvement of mechanical properties can be attributed to the increase of the cross-linkage density and aromaticity of the developed PIR foams. At an isocyanate index of 250, a drop in mechanical properties was noticed, which can be explained by the insufficient

http://dx.doi.org/10.5772/62539

technological conditions during the foaming and PIR foam curing process, such as temperature.

SEM images of the developed PU foams are shown in **Figure 13**. It is clear that EG is evenly distributed across the PU foam and it is located in walls and cell cross points. A closer look at the EG particle shows that there is no phase separation between the PU matrix and EG. It can be seen that the PU foam with EG has a much more uneven cell structure, which results in lower mechanical properties.

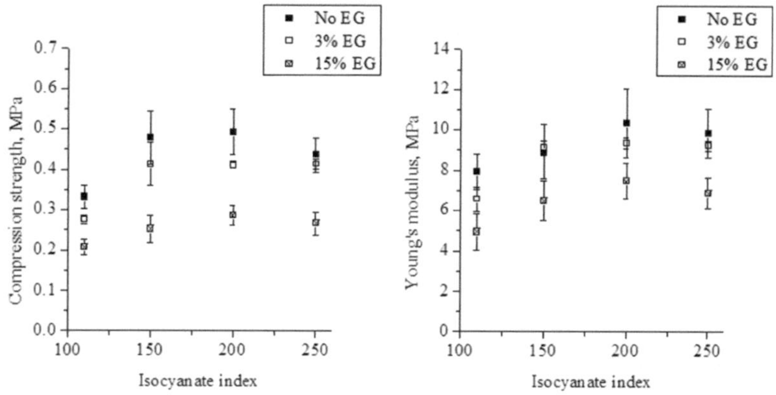

Figure 12. Compression strength and Young's modulus of rigid PU/PIR foams with different isocyanate indices and different EG content.

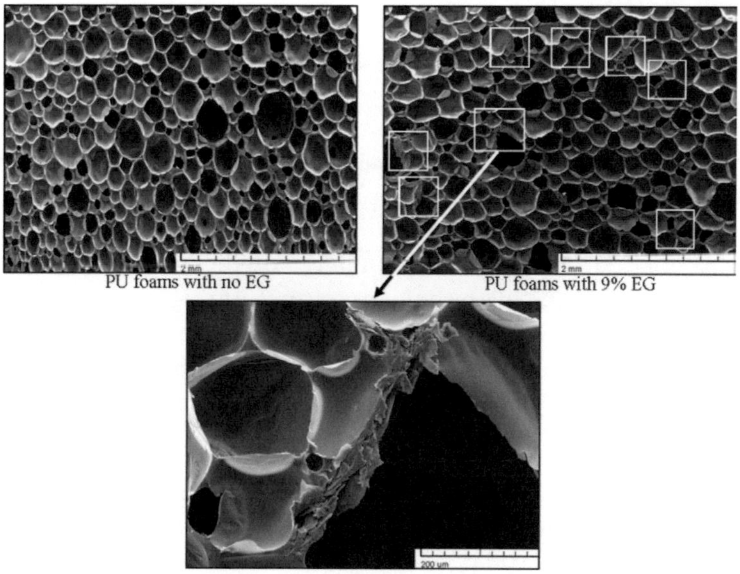

Figure 13. SEM images of PU foams with no conventional FR and 9% EG.

The main objective of this study was to obtain rigid PU/PIR foams with decreased flammability without losing superior thermal insulation properties. This was done by protecting one side of the PU foam material with IF, as shown in **Figure 14**. It is seen that the PU matrix adhesion to glass fiber is excellent and there are even cells of PU foam blown into IF. The good-phase adhesion of the PU foam and IF means that this fire protection would be easy to upscale, for example, in the production of PU/PIR sandwich-type panels. IF could replace paper roll that is used in PU/PIR sandwich-type panel production conveyor belts. This technique would introduce IF without using any adhesives that increase the production costs.

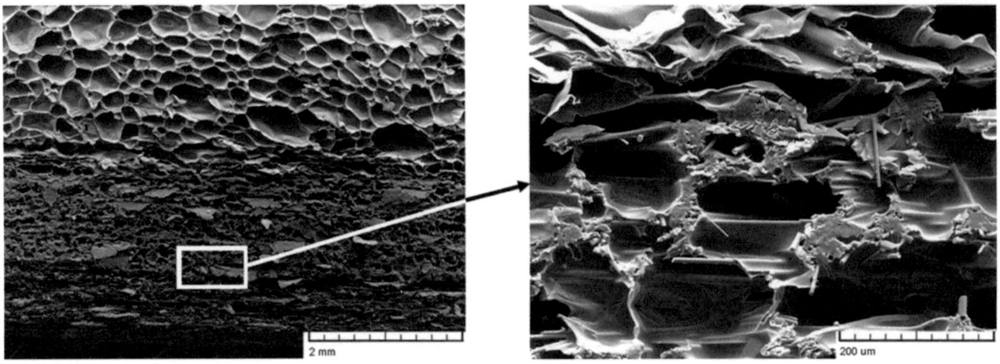

Figure 14. SEM images of PU foams with IF.

3.3. Flammability of rigid PU/PIR foams

3.3.1. EG influence on heat release of PU/PIR foams in the cone calorimeter test

The cone calorimeter test gives a significant amount of information about the flammability of the tested materials and their reaction to the heat flux. This test is a useful technique to compare different materials and gain the initial idea how a material will react in a case of fire. In the cone calorimeter test, the material is subjected to the radiant heat flux $q_{ext} = 0–100$ kW/m^2, where the flux of 35 or 50 kW/m^2 is commonly used. The small sample size of 100 mm × 100 mm × 50 mm makes this technique very useful to compare and optimize materials as for the single burning item test where much bigger samples are required. The cone calorimeter test gives the initial idea how a material will behave in an open-fire scenario. Usual parameters that are obtained after this test are HRR curves in time, time to ignition (TTI), total heat release (THR), pHRR, TSR, and MARHE.

pHRR can be used to compare materials provided that the test setup conditions are kept the same for all tested samples (sample thickness, sample holder, distance from the heater, heat flux, airflow). **Figure 15** shows a comparison of rigid PU foams from the TO polyol filled with different amounts of EG and two different liquid additive FRs (TCPP and DMPP). The addition of EG into PU foams decreased pHRR from 327.4 to 102.9 kW/m^2 when no other FR was used.

http://dx.doi.org/10.5772/62539

There was no synergistic effect between EG and DMPP or TCPP because the pHRR of PU foam samples with 15% of EG and 7% of liquid FR did not decrease below 102.9 kW/m². Although the sample with only DMPP showed the lowest values of pHRR and THR of 179.1 and 28.2 kW/m², respectively, in comparison with the sample with no FR-327.4 and 55.3 kW/m², its application could be limited due to increase of λ mentioned in **Figure 9**. DMPP works as plasticizer and promotes easier diffusion of gas from PU/PIR foam cells. The blowing agent in PU/PIR foam cells is replaced with air and λ is increased from 24.44 to 28.74 mW/m·K.

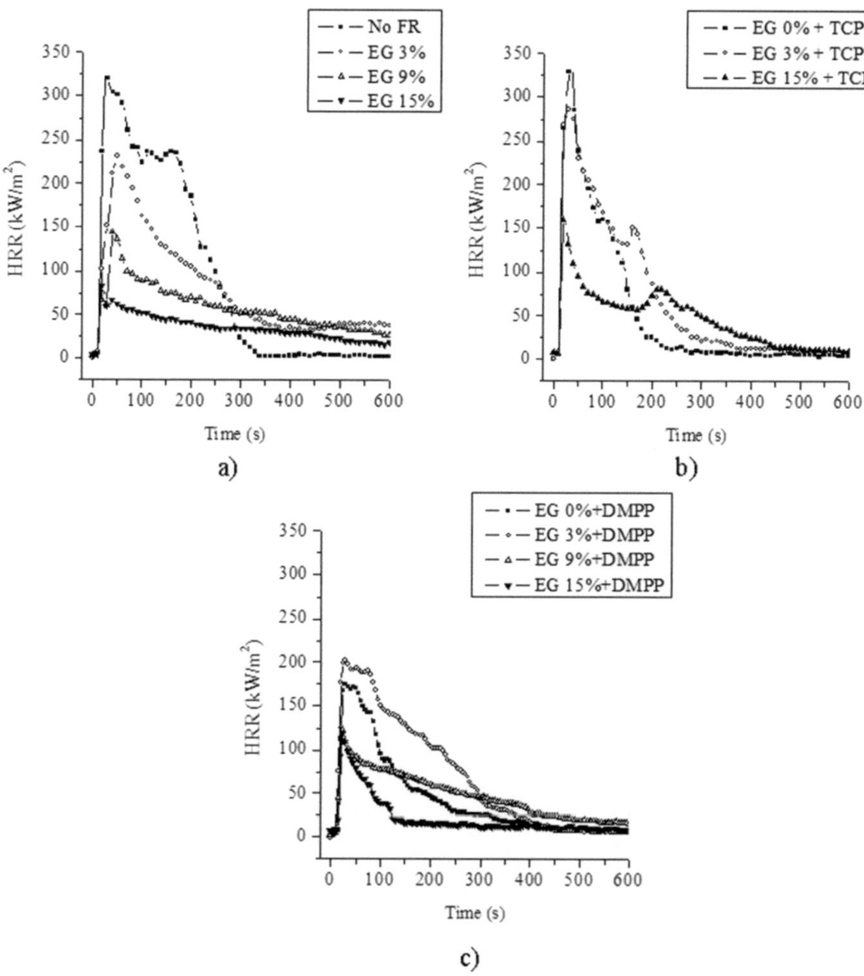

Figure 15. HRR curves of PU foam based on TO polyol with II-110, different amount of EG, and (a) no liquid FR, (b) 7% TCPP, and (c) 7% DMPP.

The addition of trimerization products of isocyanate groups into the polymer matrix increases its thermal stability and decreases flammability, which can be seen in **Figure 16(a)**. The increase of the isocyanate index from 110 to 250 decreased pHRR from 354.5 to 304.8 kW/m². The increase of the isocyanate index prolonged the burning of PU/PIR foam but the material burned with less intensity. **Figure 16(b)** shows HRR curves of PU/PIR foams protected with IF. There was no heat release detected because the material did not ignite during the whole test time. The fabric did what it was designed to do and formed a protective char layer on top of the PU/PIR foam.

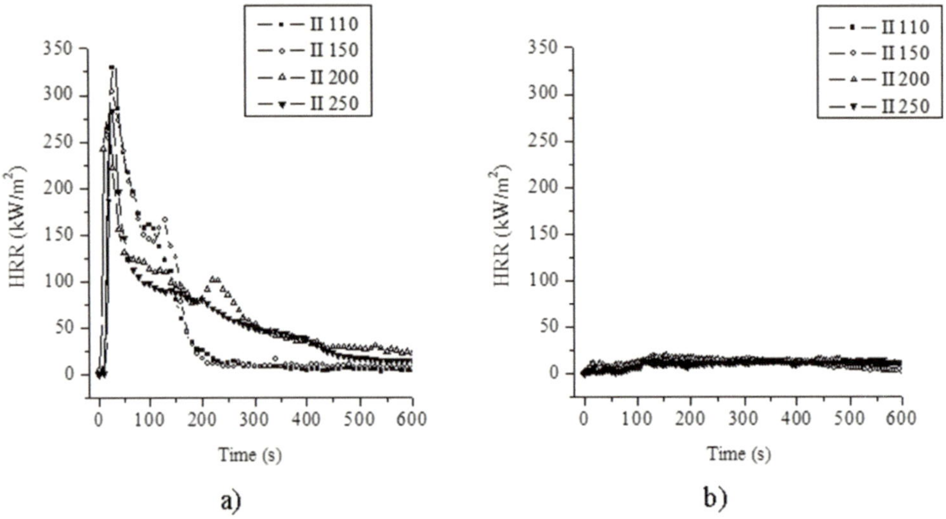

Figure 16. Rigid PU/PIR foam HRR curves: (a) foams with different isocyanate index and (b) PU/PIR foams with IF protection.

Figure 17 demonstrates the images of PU/PIR foams during the cone calorimeter test. The IF formed a thick char layer on top of the PU/PIR foam that stopped the degradation of the PU/PIR polymer matrix. IF gives excellent fire protection while keeping low thermal conductivity values as seen in **Figures 8** and **9**.

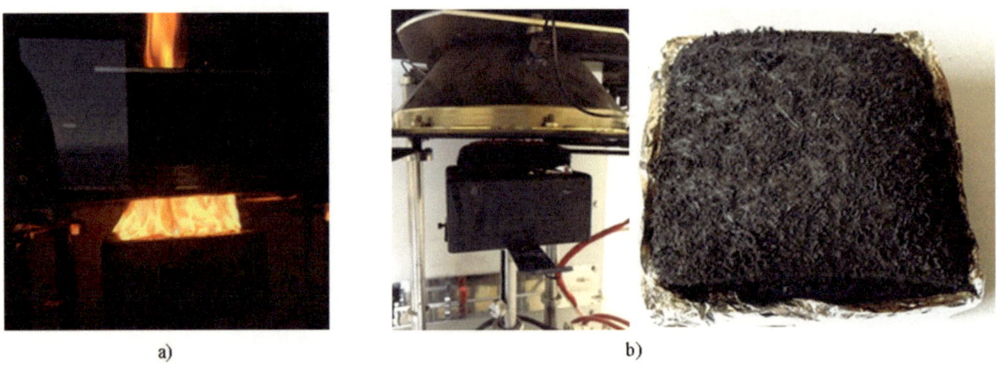

Figure 17. Rigid PU/PIR during the cone calorimeter test: (a) foams with no IF and (b) PU/PIR foams with IF protection.

Most significant flammability data after the cone calorimeter test of TO polyol-based PU foam with the isocyanate index 110 and different amounts of EG are summarized in **Figure 18**. The liquid FR TCPP showed the smallest effect on PU foam flammability when compared to DMPP at low loads of EG. However, there was a synergistic effect of TCPP together with EG, where at 9 and 15% of EG, pHRR and TSR showed lower values than the PU foam with DMPP and with no FR. A simplified interpretation of the cone calorimeter data can be done using the MARHE coefficient. This index is deduced from the maximum of HRR, which is often considered to be one of the most important fire hazards. However, this coefficient can be used to compare the materials tested under the same conditions. It can be seen that the most significant reduction of MARHE is for samples with 9 and 15% loads of EG and that only DMPP decreases MARHE for samples with no EG. In summary, the EGs significantly decrease the flammability of PU foams.

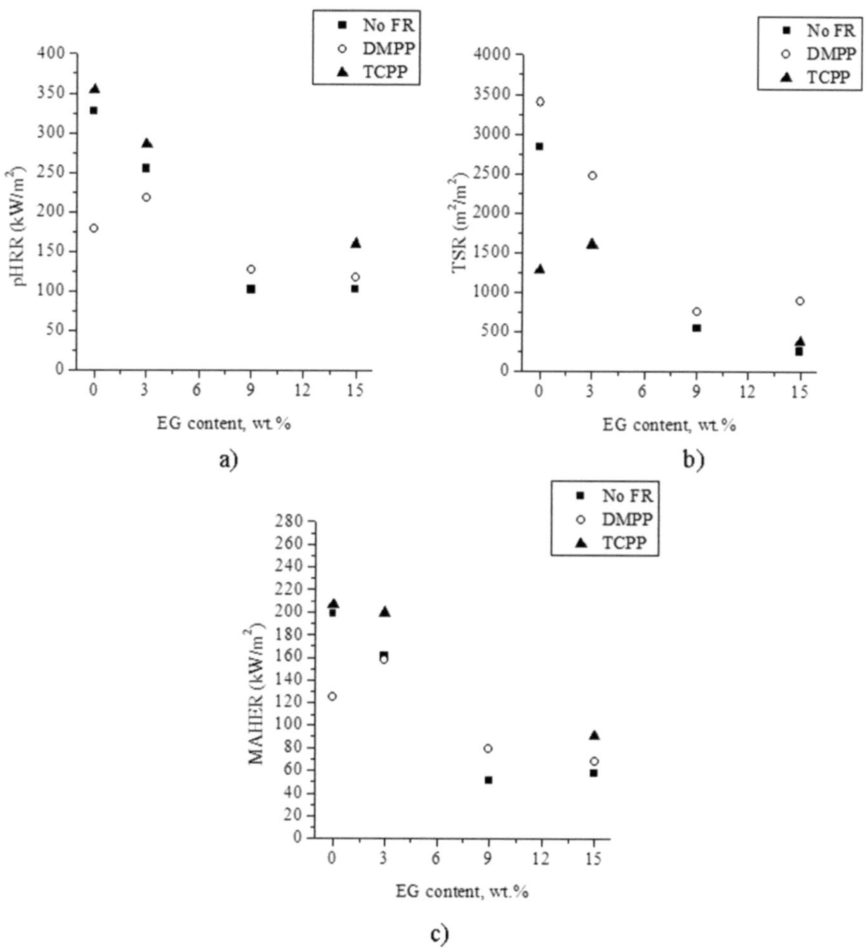

Figure 18. (a) pHRR, (b) TSR, and (c) MARHE of PU foam with different loads of EG and two different liquid FRs.

Figure 19 shows the pHRR and MARHE coefficients of rigid PU/PIR foams with different isocyanate indices and with and without IF fire protection. As mentioned, the samples with IF did not ignite, so the flammability characteristics are very low. For samples with no IF, the pHRR decreased with the increase of the isocyanate index, which correlates with the statement that cyclotrimerization groups are more thermally stable. Similarly to the case of the PU foam with an isocyanate index of 110, the addition of EG decreased PHRR and MARHE.

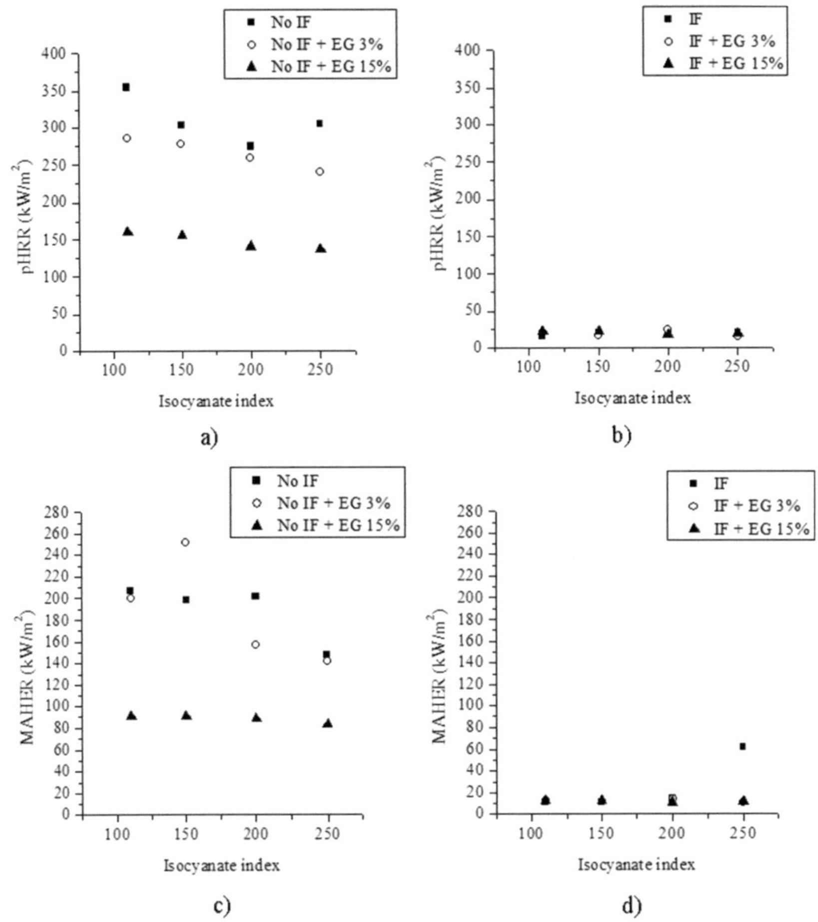

Figure 19. pHRR for rigid PU/PIR foams: (a) with no IF, (b) with IF; and MARHE: (c) with no IF, and (d) with IF.

Smoke is very important fire hazard as the majority of human victims in a fire disaster are due to poisoning with toxic gases. The cone calorimeter gives very little information about the chemical consistency of smoke. The data obtained is the solid particle amount in the smoke produced while burning the sample. **Figure 20** shows graphs with TSR data of rigid PU/PIR foams with different isocyanate indices and different EG amount. It can be seen that the addition of EG and the increase of the isocyanate index significantly decrease the amount of the produced smoke.

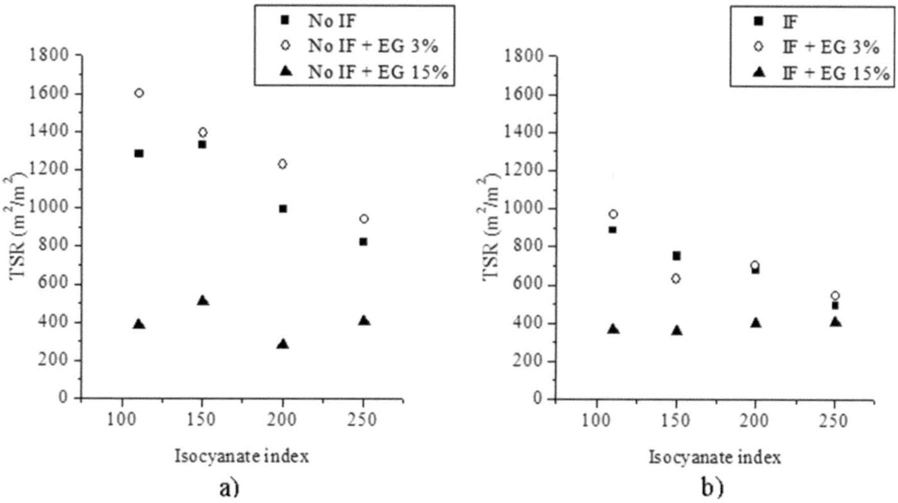

Figure 20. Total smoke released of rigid PU/PIR foams: (a) with no IF and (b) with IF.

TGA data for the developed rigid PU and PIR foams are presented in **Figure 21**. The first initial step of mass loss at 200–220°C can be attributed to the release of the additive liquid FR, unreacted isocyanate/polyols, water moisture, and other small volatiles. The next stage of degradation with the mass loss at 270°C is related to the degradation of PU soft segments in the case of the PU foam with the isocyanate index 110, as well as decoupling and pyrolysis of fatty acid dangling chains [37, 38]. At 327–351°C, breaking of isocyanate structures is observed for PU and PIR foams. Then, at 380–454°C, decomposition of aromatics and char products is seen. The rigid PIR foams with the isocyanate index of 250 produced more char residue than PU foams, namely, 20.6% and 12.3%, respectively. The TGA confirms the previous statement that PIR foams are a more thermally stable polymer material.

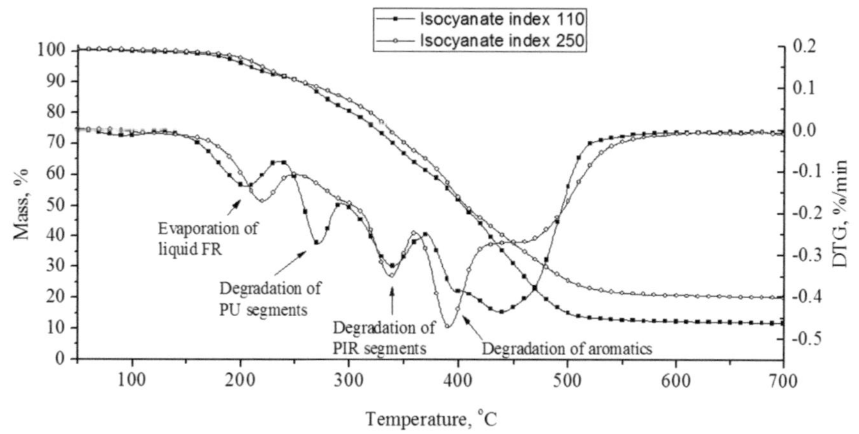

Figure 21. TGA curves for rigid PU and PIR foams obtained from TO polyol.

4. Summary

This study showed the properties of rigid PU/PIR foams obtained from different renewable raw materials—bio-based polyols synthesized from RO and TO. Bio-based polyols are a suitable substitution for petrochemical materials, as the main characteristic of a thermal insulation material—thermal conductivity—was at 22.0 mW/m·K, which is considered the industry standard for this type of material.

This study also showed a novel method of replacing liquid halogenated FR with intumescent fire protection of rigid PU/PIR foams using flexible, thermally expandable IF and EG. The addition of IF into rigid PU/PIR foams would be easily upscalable in the production of sandwich-type panels. The study of SEM images showed that no additional adhesive is needed as IF is incorporated into the PU/PIR foam structure. The synergy between two types of liquid FR (TCPP and DMPP) and additive EG was investigated but no significant improvement in fire retardancy was detected. At lower loads of EG, only DMPP showed the reduction of the heat release of PU foams. The conventional FR as well as EG increased the thermal conductivity of the developed PU/PIR foams, which gives preference to IF, where no reduction of thermal insulation properties was noticed. Also additive FR decreased the physical-mechanical properties and disrupted the morphology of PU foams. Nevertheless, one of the best fire performances of PU foam filled with 15% of EG with 7% of the conventional DMPP FR was determined. EG provided the lowest peak of heat release in the cone calorimeter test of 102.9 kW/m^2. Although EG provided excellent fire retardancy properties, the increase in thermal conductivity from 22.80 to 35.72 mW/m·K would not be acceptable for commercial use of this material. Therefore, IF is the best option as it not only stopped the material from igniting in the cone calorimeter test but also did not influence the thermal conductivity of the developed PU/PIR foams.

Acknowledgements

The authors acknowledge COST MP 1105 action "Sustainable flame retardancy for textiles and related materials based on nanoparticles substituting conventional chemicals (FLARETEX)" which supported this study and special gratitude to Technical Fibre Products Ltd. for providing free samples of two types of flexible, thermally expandable IF.

Author details

Mikelis Kirpluks*, Ugis Cabulis and Andris Avots

*Address all correspondence to: mkirpluks@gmail.com

Latvian State Institute of Wood Chemistry, Riga, Latvia

References

[1] M. Ionescu. Chemistry and Technology of Polyols for Polyurethanes Shawbury, Shrewsbury, Shropshire, United Kingdom: Rapra Technology; 2005. 586 p. ISBN 978-1-85957-501-7.

[2] M. Szycher. Szycher's Handbook of Polyurethanes. First edition. Boca Raton, USA: CRC Press; 1999. 696 p. ISBN 10: 0849306027.

[3] F. Seniha Güner, Y. Yağcı, A. T. Erciyes. Polymers from triglyceride oils. Progress in Polymer Science. 2006;31(7):633–670. DOI: 10.1016/j.progpolymsci.2006.07.001.

[4] L. Montero De Espinosa, M. A. R. Meier. Plant oils: the perfect renewable resource for polymer science?!. European Polymer Journal. 2011;47:837–852. DOI: 10.1016/j.eurpolymj.2010.11.020.

[5] M. A. Mosiewicki, M. I. Aranguren. A short review on novel biocomposites based on plant oil precursors. European Polymer Journal. 2013;49(6):1243–1256. DOI: 10.1016/j.eurpolymj.2013.02.034.

[6] J. M. Raquez, M. Deléglise, M.-F. Lacrampe, P. Krawczak. Thermosetting (bio)materials derived from renewable resources: a critical review. Progress in Polymer Science. 2010;35(4):487–509. DOI: 10.1016/j.progpolymsci.2010.01.001.

[7] C. Scrimgeour. Chemistry of fatty acids. In: F. Shahidi, editor. Bailey's Industrial Oil and Fat Products. Hoboken, USA: John Wiley & Sons, Inc.; 2005. p. 1–44. DOI: 10.1002/047167849X.bio005.

[8] Z. Petrovic. Polyurethanes from vegetable oils. Polymer Reviews. 2008;48(1):109–155. DOI: 10.1080/15583720701834224.

[9] U. Stirna, A. Fridrihsone, M. Misane, Dz. Vlsone. Rapeseed oil as renewable resource for polyol synthesis. Scientific Journal of Riga Technical University. 2011;6:85–90.

[10] M. Kirpluks, U. Cābulis, M. Kurańska, A. Prociak. Three different approaches for polyol synthesis from rapeseed oil. Key Engineering Materials. 2013;559:69–74. DOI: 10.4028/www.scientific.net/KEM.559.69.

[11] M. Kuranska, A. Prociak, M. Kirpluks, U. Cabulis. Polyurethane–polyisocyanurate foams modified with hydroxyl derivatives of rapeseed oil. Industrial Crops and Products. 2015;74:849–857. DOI: 10.1016/j.indcrop.2015.06.006.

[12] M. Zieleniewska, M. K. Leszczyński, M. Kurańska, A. Prociak, L. Szczepkowski, M. Krzyżowska, J. Ryszkowska. Preparation and characterisation of rigid polyurethane foams using a rapeseed oil-based polyol. Industrial Crops and Products. 2015;74:887–897. DOI: doi:10.1016/j.indcrop.2015.05.081.

[13] M. Kurańska, A. Prociak, M. Kirpluks, U. Cabulis. Porous polyurethane composites based on bio-components. Composites Science and Technology. 2013;74:70–76. DOI: 10.1016/j.compscitech.2012.11.014.

[14] A.S. Hasanov, H.X. Do, M. S. Shaiban. Fossil Fuel Price Uncertainty and Feedstock Edible Oil Prices: Evidence from MGARCH-M and VIRF Analysis. Energy Econ.; 2016;55:16-27. DOI: 10.1016/j.eneco.2016.04.015.

[15] R. K. Helling, D. Russell. Use of life cycle assessment to characterize the environmental impacts of polyol production options. Green Chemistry. 2009;11(3):380. DOI: 10.1039/b815833a.

[16] E. Ferri, D. Talentino. Bio-resins from cashew nutshell oil. Reinf. Plast.; 2011;55(3):29-31. DOI: 10.1016/S0034-3617(11)70074-1.

[17] L. Ugarte, A. Saralegi, R. Fernández, L. Martín, M. A. Corcuera, A. Eceiza. Flexible polyurethane foams based on 100% renewably sourced polyols. Ind. Crops Prod.; 2014;62:545-551. DOI: 10.1016/j.indcrop.2014.09.028.

[18] S. Tan, T. Abraham, D. Ference, C. W. Macosko. Rigid polyurethane foams from a soybean oil-based Polyol. Polymer. 2011;52(13):2840–2846. DOI: 10.1016/j.polymer.2011.04.040

[19] K. Hill. Fats and oils as oleochemical raw materials. Pure and Applied Chemistry. 2000;72(7):1255–1264.

[20] E. Malewska, A. Prociak. The effect of nanosilica filler on the foaming process and properties of flexible polyurethane foams obtained with rapeseed oil-based polyol. Polimery. 2015;60(7):472–479. DOI: dx.doi.org/10.14314/polimery.2015.472.

[21] S. Tan, T. Abraham, D. Ference, C. W. Masocko. Rigid polyurethane foams from a soybean oil-based polyol. Polymer. 2011;52(13):2840–2846. DOI: 10.1016/j.polymer.2011.04.040.

[22] S. N. Naik, V. V. Goud, P. K. Rout, A. K. Dalai. Production of first and second generation biofuels: a comprehensive review. Renewable and Sustainable Energy Reviews. 2010;14(2):578–597. DOI: 10.1016/j.rser.2009.10.003.

[23] R. E. H. Sims, W. Mabee, J. N. Saddler, M. Taylor. An overview of second generation biofuel technologies. Bioresource Technology. 2010;101(6):1570–1580. DOI: 10.1016/j.biortech.2009.11.046.

[24] U. Cabulis, M. Kirpluks, U. Stirna, M. J. Lopez, M. C. Vargas-Garcia. Rigid polyurethane foams obtained from tall oil and filled with natural fibers: application as a support for immobilization of lignin-degrading microorganisms. Journal of Cellular Plastics. 2012;48(6):500–515. DOI: 10.1177/0021955X12443142.

[25] K. Pietrzak, M. Kirpluks, U. Cabulis, J. Ryszkowska. Effect of the addition of tall oil-based polyols on the thermal and mechanical properties of ureaurethane elastomers.

Polymer Degradation and Stability. 2014;108:201–211. DOI: 10.1016/j.polymdegradstab.2014.03.038.

[26] A. Demirbas. Methylation of wood fatty and resin acids for production of biodiesel. Fuel. 2011;90(6):2273–2279. DOI: 10.1016/j.fuel.2011.02.037.

[27] F. Balo. Feasibility study of "Green" insulation materials including tall oil: environmental, economical and thermal properties. Energy and Buildings. 2014;86:161–175. DOI: 10.1016/j.enbuild.2014.09.027.

[28] L. Zhang, M. Zhang, Y. Zhou, L. Hu. The study of mechanical behavior and flame retardancy of castor oil phosphate-based rigid polyurethane foam composites containing expanded graphite and triethyl phosphate. Polymer Degradation and Stability. 2013;98(12):2784–2794. DOI: 10.1016/j.polymdegradstab.2013.10.015.

[29] C. F. Cullis, M. M. Hirschler. The Combustion of Organic Polymers. Oxford, United Kingdom: Clarendon Press; 1981. 419 p.

[30] B. Czupryński, J. Paciorek-Sadowska, J. Liszkowska. The effect of tri(1-chloro-3-etoxy-propane-2-ol) borate on the properties of rigid polyurethane-polyisocyanurate foams. Polimery. 2002;47(10):727–729.

[31] A. A. Lubguban, Yuan-Chan Tu, Z. R. Lozada, Fu-Hung Hsieh, G. J. Suppes. Functionalization via glycerol transesterification of polymerized soybean oil. Applied Polymer. 2009;112(1):19–27. DOI: 10.1002/app.29382.

[32] M. Modesti, A. Lorenzetti, F. Simioni, M. Checchin. Influence of different flame retardants on fire behaviour of modified PIR/PUR polymers. Polymer Degradation Stability. 2001;74(3):475–479. DOI: 10.1016/S0141-3910(01)00171-9.

[33] U. Stirna, U. Cabulis, I. Beverte. Water-blown polyisocyanurate foams from vegetable oil polyols. Journal of Cellular Plastics. 2008;44(2):139–160. DOI: 10.1177/0021955X07084705.

[34] B. Schartel, T. Hull. Development of fire-retarded materials—interpretation of cone calorimeter data. Fire and Materials. 2007;31(5):327–354. DOI: 10.1002/fam.949.

[35] X.-C. Bian, J.-H. Tang, Z.-M. Li, Z.-Y. Lu, A. Lu. Dependence of flame-retardant properties on density of expandable graphite filled rigid polyurethane foam. Journal of Applied Polymer Science. 2007;104(5):3347–3355.

[36] X.-M. Hu, D.-M. Wang. Enhanced fire behavior of rigid polyurethane foam by intumescent flame retardants. Journal of Applied Polymer Science. 2012;129(1):238–246. DOI: 10.1002/app.38722.

[37] L. Gao, G. Zheng, Y. Zhou, L. Hu, G. Feng, Y. Xie. Synergistic effect of expandable graphite, melamine polyphosphate and layered double hydroxide on improving the fire behavior of rosin-based rigid polyurethane foam. Industrial Crops and Products. 2013;50:638–647. DOI: 10.1016/j.indcrop.2013.07.050.

[38] X. Hua, D. Wang, S. Wang. Synergistic effects of expandable graphite and dimethyl methyl phosphonate on the mechanical properties, fire behavior, and thermal stability of a polyisocyanurate–polyurethane foam. International Journal of Mining Science and Technology. 2013;23(1):13–20. DOI: 10.1016/j.ijmst.2013.01.003.

[39] X.-C. Bian, J.-H. Tang, Z.-M. Li. Flame retardancy of hollow glass microsphere/rigid polyurethane foams in the presence of expandable graphite. Journal of Applied Polymer Science. 2008;109(3):1935–1943. DOI: 10.1002/app.27786.

[40] D. K. Chattopadhyay, D. C. Webster. Thermal stability and flame retardancy of polyurethanes. Progress in Polymer Science. 2009;34(10):1068–1133. DOI: 10.1016/j.progpolymsci.2009.06.002.

[41] M. Thirumal, D. Khastgir, G. B. Nando, Y. P. Naik, N. K. Singha. Halogen-free flame retardant PUF: effect of melamine compounds on mechanical, thermal and flame retardant properties. Polymer Degradation and Stability. 2010;95(6):1138–1145. DOI: 10.1016/j.polymdegradstab.2010.01.035.

[42] M. Thirumal, D. Khastgir, N. K. Singha, B. S. Manjunath, Y. P. Naik. Effect of expandable graphite on the properties of intumescent flame-retardant polyurethane foam. Journal of Applied Polymer Science. 2008;110(5):2586–2594. DOI: 10.1002/app.28763.

[43] X.-G. Zhang, L.-L. Ge, W.-Q. Zhang, J.-H. Tang, L. Ye, Z.-M. Li. Expandable graphite-methyl methacrylate-acrylic acid copolymer composite particles as a flame retardant of rigid polyurethane foam. Journal of Applied Polymer Science. 2011;122(2):932–941. DOI: 10.1002/app.34198.

[44] S. Duquesne, M. Bras, S. Bourbigot, R. Delobel, H. Vezin, G. Camino, B. Eling, C. Lindsay, T. Roels. Expandable graphite: a fire retardant additive for polyurethane coatings. Fire and Materials. 2003;27(3):103–117. DOI: 10.1002/fam.812.

[45] M. Modesti, A. Lorenzetti, F. Simioni, G. Camino. Expandable graphite as an intumescent flame retardant in polyisocyanurate–polyurethane foams. Polymer Degradation and Stability. 2002;77(2):195–202. DOI: 10.1016/S0141-3910(02)00034-4.

[46] L. Shi, Z.-M. Li, B.-H. Xie, J.-H. Wang, C.-R. Tian, M.-B. Yang. Flame retardancy of different-sized expandable graphite particles for high-density rigid polyurethane foams. Polymer International. 2006;55(8):862–871. DOI: 10.1002/pi.2021.

[47] M. Kirpluks, U. Cabulis, V. Zeltins, L. Stiebra, A. Avots. Rigid polyurethane foam thermal insulation protected with mineral intumescent mat. Autex Research Journal. 2014;14(4):259–269. DOI: 10.2478/aut-2014-0026 © AUTEX.

[48] D. Price, Yan Liu, T. R. Hull, G. J. Milnes, B. K. Kandola, A. R. Horrocks. Burning behaviour of foam/cotton fabric combinations in the cone calorimeter. Polymer Degradation and Stability. 2002;77(2):213–220. DOI: 10.1016/S0141-3910(02)00036-8.

[49] T. M. Kotresh, R. Indushekar, M. S. Subbulakshmi, S. N. Vijayalakshmi, A. S. Krishna Prasad, K. Gaurav. Evaluation of foam/single and multiple layer Nomex fabric

combinations in the cone calorimeter. Polymer Testing. 2005;24(5):207–612. DOI: 10.1016/j.polymertesting.2005.03.001.

[50] S. Vijayakumar, P. R. Rajakumar. Infrared spectral analysis of waste pet samples. International Letters of Chemistry, Physics and Astronomy. 2012;4:58–65. ISSN: 2299-3843.

[51] E. Žagar, J. Grdadolnik. An infrared spectroscopic study of H-bond network in hyperbranched polyester polyol. Journal of Molecular Structure. 2003;658(3):143–152. DOI: 10.1016/S0022-2860(03)00286-2.

[52] Wu Lifeng, J. Gemert, R. E. Camargo. Rheology Study in Polyurethane Rigid Foams. Auburn Hills, USA: Huntsman Corporation; 2008.

[53] A. Prociak, G. Rokicki, J. Ryszkowska. Materialy poliuretanowe. Warsaw, Poland: Wydawnictwo Naukowe PWN; 2014. 404 p. ISBN:9788301174187.

Chapter 6

The Review of Some Commonly Used Methods and Techniques to Measure the Thermal Conductivity of Insulation Materials

Numan Yüksel

Additional information is available at the end of the chapter

http://dx.doi.org/10.5772/64157

Abstract

The use of insulation materials is considered as one of the most effective means of conserving energy in various fields. Thermal insulation materials enable systems to achieve energy efficiency. Many different thermal insulation materials have been developed to reduce heat flow by limiting conduction, convection, and/or radiation while performing one or more functions. These functions may vary in the context of thermal design, numerical simulations, and a wide range of engineering problems, such as determining the heat loss, temperature field, isolation, and cooling conservation, and in a variety of other technologies. One of the most effective ways to identify and determine performance is effective thermal conductivity. The thermal measurement performance is usually evaluated in a temperature and signal gradient for single or combined homogeneous/heterogenous materials. The two main categories of thermal conductivity measurement techniques are steady-state methods and transient methods. The aim of this chapter is to present various measurement methods and to investigate their suitability for method purposes. This chapter presents new and accurate experimental techniques and methods for measuring the thermal conductivity of several most commonly used insulation materials. Some of these methods are commonly used in the field for measuring the thermal property of insulation materials. On the other hand, different insulation measurement practices are presented depending upon the overall structures. The analysis predicting the thermal conductivities of insulation materials is also discussed.

Keywords: heat transfer, insulation material, mathematical model, measurement method, tests, thermal conductivity

INTECH
open science | open minds

1. Introduction

The development of technologies leads to changes in demand. Insulation materials are created in several forms including porous form, blanket or batt form, rigid form, natural form, foamed structure, and reflective structure. Fiber and polymer products are the most commonly used types of thermal insulation. Many studies have examined the impact of various parameters on the thermal performance of insulation materials. Extensive investigations have focused on heat transfer in these materials in the context of their numerous and varied applications. The thermal conductivity in these applications is one of the most important challenges facing thermal, mechanical, material, and civil engineers. In various fields, the accuracy of different techniques for evaluating thermal conductivity and other properties is widely debated as a fundamental parameter. As a result of the wide range of thermal properties of insulation materials, there is no single measurement method for all thermal conductivity measurements [1].

The accuracy of manufacturers' claimed values for thermal properties is sometimes questionable since the thermal data for certain materials are often incomplete and lacking in important information. When values for insulation types are quoted, manufacturers do not always report the density and temperature of the materials tested. In general, the "effective" thermal conductivities of the materials depend on the constituents and/or voids present in the different characteristics of their structures as well as the density and temperature of the material.

In the field of thermal insulation technology, many developments during the last two decades have enhanced the accuracy of measurement techniques as well as the current understanding of the principles of heat transfer through different materials. These techniques are thus differentiated mainly by the range of thermal conductivity, the range of material types, measurement time, measurement accuracy, specimen type, and temperature range.

In the study, a survey concerning available experimental techniques for the measurements is conducted. The main focus is on descriptive measurement methods, and their ranges of the thermal conductivity and temperature are determined. The objective is to analyze a measurement apparatus designed to determine effective thermal conductivity of insulation materials. The other aim of this chapter is to figure out the models for the effective thermal conductivity of insulation materials. The prediction of the property has been determined using experimental and analytical models in different studies. The accuracy of any method and model is limited by physical properties and other factors. However, measurement and modeling of thermal conductivity are difficult and require high precision in the determination of the various parameters involved in the calculations. To analyze thermal behavior of materials, the methods and the models must be clearly known and defined.

Insulation materials such as natural or man-made materials differ concerning the material structure and the range of use. To develop insulation materials in an economically and environmentally friendly manner, it is essential to have knowledge and control over their thermal conductivities. The properties can vary with temperature, pressure, and composition, affecting the transfer of heat. To answer the following questions, thermal conductivity must be known [2].

- How is a particular insulation material performing?
- How does performance change with the weather and different conditions, and how can the performance be improved?
- What is the optimum insulation for technologies/systems operating under different temperature, gas, or pressure conditions?
- How can a system be designed to achieve the required efficiency, and what are the best materials to use?
- What is the heating/cooling load of a building and a structure?
- How can cryogenic tanks be insulated in the best possible way?
- How can the heat transfer from an electronic component be improved?

There are a number of methods to measure thermal conductivity. In general, there are two basic techniques for measuring thermal conductivity: steady-state methods and transient or non-steady-state methods [1, 3]. Each of these methods is suitable for a limited range of materials, and they are based on the fundamental laws of heat conduction and electrical analogy. Steady-state methods have been traditionally used since they are mathematically simpler. There is an important distinction between steady-state and transient techniques [4, 5]. Transient heat transfer methods are capable of directly determining thermal diffusivity, whereas steady-state methods are considered to be more accurate than transient methods for testing dry materials [6].

The **steady-state technique** records a measurement when a tested material's thermal state reaches complete equilibrium [5]. A steady-state condition is attained when the temperature at each point of the specimen is constant and the temperature does not change with time. A disadvantage, however, is that it generally takes a long time to reach the required equilibrium [4, 5]. The method involves expensive method apparatus since a well-designed experimental installation system is usually needed. Nevertheless, it is the primary and most accurate measurement method.

The **non-steady-state or transient** technique records a measurement during the heating process. The method determines thermal conductivity properties by means of transient sensors. These measurements can be made relatively quickly, which garners an advantage over steady-state techniques [4, 5, 7]. For this reason, numerous solutions have been derived for the transient heat conduction equation by using one-, two-, three-dimensional geometries [7]. Transient methods generally employ needle probes or wires [4].

Compared to electrical and thermal transport, the ratios of thermal conductivities of the best conduction and insulation conditions are significative and determinative magnitudes. Therefore, instruments for thermal property identification are often designed only for specific kinds of materials or temperature ranges. **Table 1** presents a comparison of the most common methods of thermal conductivity measurement [7]. Measurement systems can also be divided into three categories based on the operating temperature of the apparatus: (1) room temperature operation (20–25°C), (2) below room temperature operation (down to about –180°C), and

(3) high-temperature operation (up to 600°C or above) [8]. A given measurement system is often optimized for one of these temperature ranges.

Method		Temperature range	Uncertainly	Materials	Positive	Negative
Steady-state methods	Guarded hot plate	80–800 K [7], −180–1000°C [9], 80–1500 K [10]	2% [7] and 0.0001–2 W/(m K) [9] 2–5% and 0.0001–1 W/(m K) [10]	Insulation Materials [7] and solid, opaque, insulators [10]	High accuracy	Long measurement, large specimen size, low conductivity materials
	Cylinder	4–1000 K [7]	2% [7]	Metals [7]	Temperature range, simultaneous determination of electrical conductivity	Long measurement
	Heat-flow meter	−100–200°C for normal [7] 90–1300 K for axial heat flow and 298–2600 K for radial heat flow [10]	3–10% [7] 0.007–1.0 W/(m K) [9] 0.5–2% and 10–500 W/(m K) (axial) and 3–15% and 0.01–200 W/(m K) (radial) [10]	Insulations, plastics, glasses, ceramics [7] Some metals, rocks, polymers [9] Metals and Solids [10]	Simple construction and operation	Measurement uncertainly, relative measurement
	Comparative	20–1300°C [7] 0–1000°C [9]	10–20% [7] and 0.2–200 W/(m K) [9]	Metals, ceramics, plastics [7]	Simple construction and operation	Measurement uncertainly, relative measurement
	Direct heating	400–3000 K [7, 10]	2–10% [7] 2–5% [10] and 10–200 W/(m K) [10]	Metals [7] Wires, rods, tubes of electrical conductors [10]	Simple and fast measurements, simultaneous determination of electrical conductivity	Only electrically conducting materials
	Pipe method	20–2500°C [7] and 50–800°C [9]	3–20% [7] and 0.02–2 W/(m K) [9]	Solids [7] calcium silicates,	Temperature range	Specimen preparation, long measurement

Method		Temperature range	Uncertainly	Materials	Positive	Negative
				mineral and refractory fiber blankets) [9]		time
Transient Methods	Hot wire, hot strip	20–2000°C [7], −40–1600°C for hot wire and −50 to 500°C for hot strip [9] 298–1800 K [10] for hot wire	1–10 % [7] 0.001–20 W/(m K) for hotwire and 0.1–5 W/(m K) for hot strip [9] 5–15% [10] and 0.02–2 W/(m K) for hotwire [10]	Liquids, glasses, solids of k_{low} [7] refractory materials [9, 10] and plastics, granules, powders for hotwire [9] and glasses, foods, ceramics for hot strip [9]	Temperature range, fast, accuracy	Limited to low conductivity materials
	Hot-disk (TPS technique)	30–1200 K [7]	-	liquids, pastes, solids and powders	very short time accuracy, different thermal properties simultaneously	the range of 0.005 and 500 W/(m K) (conducting or insulating material)
	Laser flash	−100–3000°C [7] and 100–3300 K [10]	3–5% [7] 1.5–5 % [10] and 0.1–1500 W/(m K) [9, 10]	solids, liquids, and powders [7] and liquid metals, polymer, ceramics [9, 10]	Temperature range, most small specimen, fast, accuracy at high temperature	Expensive, not for insulation materials
	Photothermal (PT), Photoacoustic	30–1500 K [7] −50–500°C [9] and 200–800 K [10] for PT	Not sufficiently known [7], 1–10 % [10] 0.1–200 W/(m K) for PT [9, 10]	Solids, liquids, gases, thin films [7], small parts of most solid [9,10]	Usable for thin films, liquids, and gases	nonstandard, knowledge about accuracy

Table 1. Comparison of measurement methods for detecting the thermal conductivity [7, 9, 10]

For measuring thermal conductivity, there are four main types of measurement setups: the guarded hot plate (GHP), the heat-flow meter (HFM), the hot wire, and laser flash diffusivity. The usage of these tools/methods differs in technique, material type, intended specimen size, measurement time, capability, and measurement methodology [5].

To analyze heat transfer behavior of insulations, a guarded hot plate or a heat-flow meter is usually used. The hot-wire and flash method use special devices for consolidated insulation specimens. The laser flash method is often employed for highly conductive ceramics, metals, and some composites [2]. The thermal conductivity of large specimens of refractory material is measured by using hot-wire systems [2]. **Figure 1** provides a comparison of measurement methods and material types for the ranges of thermal conductivity [2].

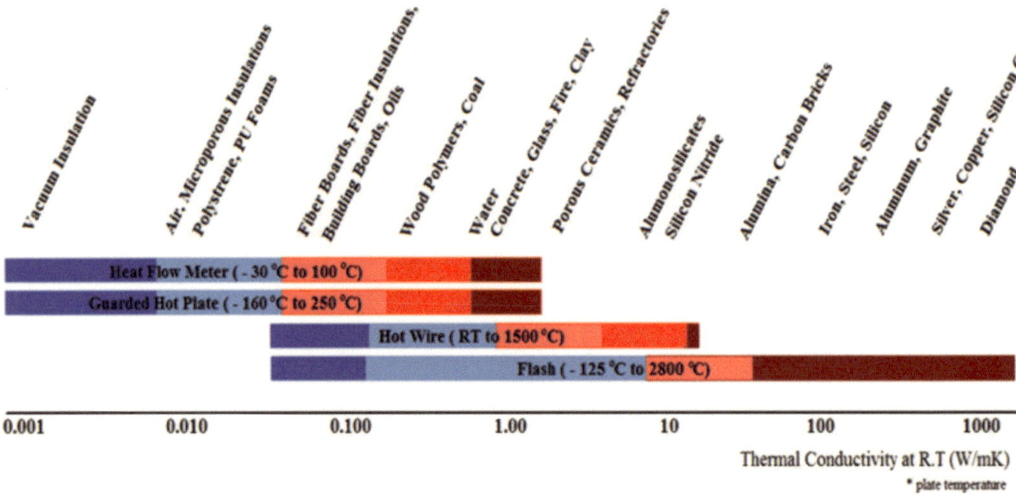

Figure 1. Comparison of measurement methods and material type for the ranges of thermal conductivity [2].

2. Steady-state methods

Steady-state methods apply Fourier's law of heat conduction to measure thermal conductivity. The solution to the problems with the different steady heat-flow methods is to convert the heat transfer problem to a one-dimensional problem, thus simplifying the mathematics. The calculations change for the models of an infinite slab, an infinite cylinder, or a sphere. The typical specimen geometry, the configuration of a measurement system, and the magnitude of the thermal conductivity are used to distinguish between different types of thermal conductivity measurements. The thermal magnitude of the measuring object is determined by the following measuring techniques using the direction of the heat flow, the conservation of the heat flow, and an auxiliary layer having a known thermal property.

2.1. Guarded hot plate

The guarded hot plate, also known as the Poensgen apparatus [11], is the most commonly used and most effective method for measuring the thermal conductivity of insulation materials. The

GHP relies on a steady temperature difference over a known thickness of a specimen and its primary purpose is to control the heat flow through the material. One disadvantage is that establishing a steady-state temperature gradient through a specimen is time-consuming when using the GHP and other steady-state techniques. Other potential disadvantages are that the temperature gradient must be relatively large, the specimen width must be large, and also that the contact resistance between the thermocouple and the specimen surface poses a major source of error [12]. Although Reference [12] cites large specimen size as a potential disadvantage, size is usually not a serious issue [8].

The experimental setup of the guarded hot plate employs a steady-state heat transfer between a hot plate and a cold plate. However, the accuracy of this method is questionable, interlaboratory comparisons of GHP calculations have revealed discrepancies among 20 different GHPs used at different times [5]. The individual results of these 20 GHPs diverged significantly from reference values, ranging from +13 to −16% [5, 11].

Despite these disadvantages, the standardized GHP method is the ideal apparatus for researchers and scientists in the field of insulation testing and it is considered an absolute measurement method. The practical applicability requires careful consideration of the array content: (a) attaining steady-state conditions; (b) the unidirectional heat flow in the area under analysis, the temperatures of the hot and cold surfaces, and the specimens' thickness; and (c) other factors influencing the unidirectional heat flow [8].

Another advantage is that the GHP method is standardized in countries such as the United States (ASTM C 177-63), Great Britain (B.S. 874:1965), and Germany (DIN 52612) [13]. The details of this method are provided by the American Society for Testing Material (ASTM) Standards associated with the method and/or materials [1]. The details of this standard are partly based on the difficulty of attaining steady-state conditions [11], accurately adjusting the temperatures in conventional plates (guarded, hot, and cold), and design conditions.

2.1.1. The construction of the guarded hot plate

The guarded hot-plate measurements are analyzed on the fundamental of the heat transfer in the infinite slab geometry. Since specimen dimensions are finite, unidirectional heat flow is achieved through the use of guard heaters. The temperature of a thermal guard is maintained at the same temperature as its adjacent surface (which is considered as an auxiliary heater/heat sink), in order to prevent heat loss from the specimen and heat source/heat sink, and as a result, unidirectional heat flow is attained [1]. After a steady state is reached, the heating and cooling plates have stable temperatures. Then the thermal conductivity can be determined based on the heat input, the temperature difference through the specimen, the thickness of the specimen, and the size of the metered area of heat transfer. Steady-state conditions may change with respect to specimen type, specimen size, and mean temperature [14]. The GHP is most suitable for dry homogeneous specimens [15], but it is unsuitable for materials in which there is a potentiality for moisture migration [16].

The guarded hot-plate setup is comprised of cold plates, a hot plate, a system of guard heaters, and thermal insulation. Hot plate is electrically heated and the cold plates are Peltier coolers or liquid-cooled heat sinks. The configuration is arranged symmetrically, with guarded hot plates located on the sides while the heater unit is sandwiched between two specimens or a single specimen and an auxiliary layer (**Figure 2**). The different types of guarded hot-plate apparatus are shown in **Figure 2**. In the single-sided system state, the heat flow passes through one specimen, while the top of the main heater acts as an insulating guard, thus ensuring an adiabatic environment [8].

In the two specimen apparatus, the main advantage is that heat loss from the hot plate can be controlled more effectively due to the symmetrical arrangement of the specimen on each side of the heater. Unlike the single-specimen method, the symmetrical setup can be used for investigating solid materials. For measuring the conductivity of nonsolid materials, it is necessary to heat the specimen from the top in order to avoid convection [7].

The electrical heating is placed into the plates in a certain shape or form, such as a square or a circular shape. The guarded plate (ring), the central plate (metered area), and the auxiliary heater can all be arranged in this manner. The apparatus must test two specimens simultaneously in the form of a slab with a standard size (such as 300 mm × 300 mm or different sizes). A fixed heat rate must be applied by an electric heater. This arrangement produces a heat flow across the two specimens, flowing outward toward two plates chilled by a Peltier or a liquid cooling system.

These heat measurements are recorded by differential thermocouples, which are instruments that control a flat electrically heated metering area that is surrounded on all lateral sides by a guard heater section. The heated section provides the planar heat source applied to the hot face of the specimens [8]. Heat is supplied to the metered area (the central heater) at an assigned heat power rate. The temperature of the guard heater is maintained at the same temperature as the metered section by using a control system. The adjacent thermal guard surfaces and/or plates are held at the same temperature range, and ideally no heat leakage occurs from the source, the specimen, or the boundaries. This is aimed at ensuring a one-dimensional thermal

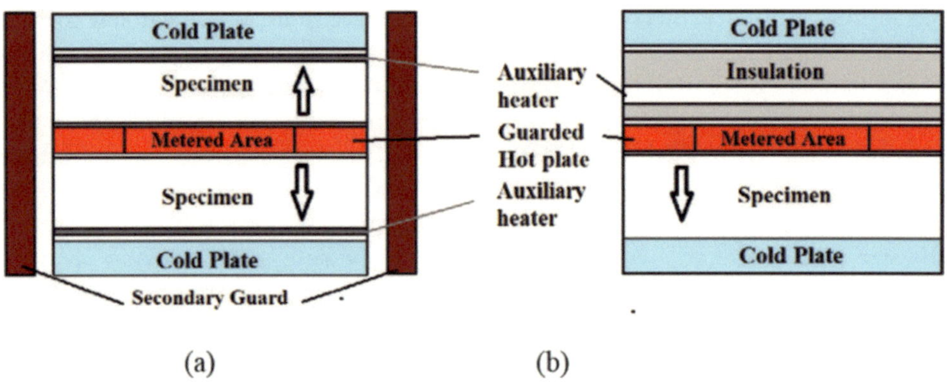

Figure 2. The apparatus of guarded hot-plate method for thermal conductivity measurement. (a) two specimen with/without auxiliary heaters and secondary guards, (b) single-specimen [7].

heat flow in the actual and practical test section, corresponding solely to the central metered heater. In addition, the apparatus is surrounded by thermal insulation, as well as guard heaters. And, the hot/cold metal parts are positioned between the heaters/cooling plates and each specimen. The parts matched the same frame design are adjacent to the related side (hot or cold) temperature sensors [13]. A data acquisition system is connected to the temperature sensors and the electrical power supply devices, which are in turn controlled by a closed-loop control system.

2.1.2. Principle of operation

The specimens of the homogeneous material with the same thickness are interposed between the hot guard heaters and the cold plates. For two specimen apparatuses, the auxiliary heaters may be placed above and below the specimens. A well-defined, user-selectable temperature difference is established between the hot and the cold plates. The power rate input in the hot plate with metered area A is measured when thermal equilibrium is reached at steady-state conditions. When the control system is used, the plate temperatures reach stability.

It is assumed that the measured heat power rate is transferred across the specimen due to guarded heaters. After thermal equilibrium has developed and the heating and cooling plates are kept in stable temperatures, the thermal conductivity can be calculated from the input values. The input values are the heat power Q, the temperature differential across the specimen ($T_{hot} - T_{cold}$), the specimen thickness (Δx), and the heat transfer area (center metered area, A). The thermal conductivity is computed by measuring the quantity of heat input under the steady-state temperature profile in the entire specimen [1, 3, 14]. From the measured input values, the effective thermal conductivity can be calculated using the following unidirectional steady-state heat transfer equation:

$$k_{eff} = \frac{Q}{2} \frac{\Delta x}{A.\Delta T} \tag{1}$$

where the heat flow Q is obtained by measuring a power P (or half power for two specimen) generated in an electrical heater. The heat conduction equation for homogenous isotropic materials without using internal heat generation is given for the steady state in Eq. (1). These methods depend on Fourier-Biot law of heat conduction [1, 3, 14]. Its modified equation forms can be used for one-dimensional steady heat flow across different sizes, such as plate, cylinder, and sphere.

For the shapes in cylinder forms, radial heat-flow steady-state methods are observed. The specimen completely encloses the heating source in this method, eliminating end losses. The lateral effects are assumed to be insignificant either because the ratio of length to diameter of the test apparatus is large or because guard heaters are used. It is assumed that the surface of the central heater at a diameter r_1 and the outer specimen surface at diameter r_2 reach the same temperature after the steady state is established. The thermal conductivity can be determined based on "the heating power, the length of the cylinder, the temperature differential between two internally located sensors, and their radial position" [16]. Because of practical application difficulties, the cylinder (and sphere) method is not popular. Nevertheless, this method is applied and used to measure thermal conductivity using the cylinder shape method.

The guarded hot-plate method under a vacuum is based on an absolute measurement method for research and therefore requires no calibration standards. Furthermore, this can be seen as an absolute measurement, regardless of vacuum conditions. The plate system is placed in a vacuum medium. The measurements can be carried out under a vacuum as well as under atmospheric or defined pressure levels. The system requires a symmetry and two specimens for each test. With a guarded heater and/or thermal insulation, a relative uncertainty of 2% for thermal conductivity measurements of can be achieved. Each plate and the guard ring/heater are connected to a separate control system with temperature sensor(s) and an assigned power supply.

The guarded hot-plate and cylinder method exemplify a measurement principle that has been optimized for different ranges of thermal conductivity. The guarded hot-plate method can be used to test the thermal properties of nonmetals such as thermal insulation materials, polymers glasses, and ceramics, as well as liquids and gases in the temperature range between about 80 and 800 K [7]. The thermal conductivities of metals (approximately up to 500 W/(m K) in a temperature range between about 4 and 1000 K) can be tested via the cylinder method of employing axial heat flow. The GHP method is appropriate for these kinds of metal because the determination of the temperature difference is the main challenge when measuring materials with high thermal conductivity (e.g., metals). In these kinds of tests, the contact resistances between the specimen and the heater or the cold plate must be considered [7].

2.2. Heat-flow meters

The use of heat-flow meters is based on substantially the same principles as other measurement techniques, but it is not identical to them [13, 17, 18]. The main disadvantage of the guarded hot-plate methods is that they are very time consuming, as stated above. In contrast, heat-flow meters are accurate and fast apparatuses, and the operation of these apparatuses is easy for measuring the thermal conductivity of low-conductivity materials [2]. The method is based on the improvement in accuracy and speed of the measurement. The maximum temperature limits are approximately 200°C for the heat-flow meter method [7] and about 100°C in practical applications [2].

The heat-flow meter is described in various standards for tests. The heat-flow meter design resembles to setup of the single-specimen guarded hot-plate apparatus. The basic idea of the heat-flow meter is deducing the heat flux based on the measurement of a drop in temperature throughout a thermal resistor. The way for heat flux measurement is carried out either by using a certified well-known reference specimen or a heat flux sensor.

The specimen is placed between two plates held in different temperatures, with one being heated and the other plate being cooled, as shown in **Figure 3.** Instead of using a main heater as in the guarded hot-plate method, heat flux transducers are used to measure the heat flow through the specimen. The heat flux is determined in a current with the measurement of a voltage drop through an electrical resistor. Sensors provide an electrical output signal. The measured signal and the change in thermovoltage is proportional to the drop in temperature drop occurring throughout the plate.

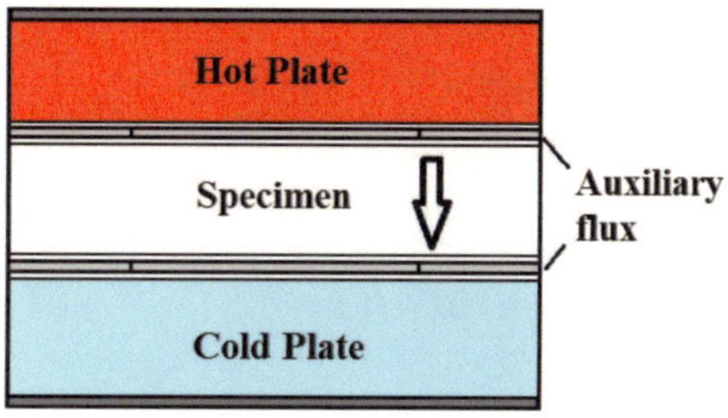

Figure 3. Schematic design of heat-flow meter basically.

Heat flux sensors are most often composed of a series of connections of thermocouples spanning a thermal resistor, e.g., a thin ceramic or plastic plate [7]. A second heat flux is occasionally applied at the cold plate is in order to measure radial heat loss and also to reduce the time needed for the measurement. This reduction in time poses an advantage for this method when measuring of insulation materials. The temperature of the plates is measured and adjusted to the desired set point when reaching a constant value. Steady-state conditions occur when the amount of heat flux is equal at each point of the layered system. After thermal equilibrium is established, the test is determined under the conditions. In order to calculate thermal conductivity, the steady-state temperatures, the specimen's thickness, the specimen's metered area, and the heat flux input to the hot plate are used. The heat flux output is usually calibrated with various reference standards, e.g., in a guarded hot-plate apparatus.

Compared to the guarded hot-plate method, which is an absolute method, the heat-flow meter method is comparative and thus can be considered to be a relative method. Insulation materials and polymers ($k < 0.3$ W/(m K)) are usually tested via the heat-flow meter method, and sometimes it is used for glasses and ceramics and for other materials with thermal conductivities lower than about 5 W/(m K) [7]. For insulation materials at about room temperature, the measurement uncertainties are approximately 3%, and at high temperatures the uncertainties are between 10 and 20% [7].

Steady-state stability can be reached in short times, resulting in productivity gains. The stable case, however, is valuable for providing repeatability, the long-term consistency of a specimen.

2.3. Direct heating method

Two disadvantages of steady-state methods are the lengthy time requirements and the difficulty of determining heat loss, especially at high temperatures. These disadvantages can be overcome by the direct heating method, which can be used for electrically conductive materials such as metals.

The specimen, such as a wire, pipe, or rod, is placed in a vacuum chamber, clamped between two heat sinks cooled by liquid, and the specimen is heated up to temperatures in the range of 300–4000 K [7]. **Figure 4(a)** portrays the schematic of the design of the direct heating method. Voltage drops and temperatures are measured: in the middle of the rod and on each end of the rod. From these three measurements obtained in the direct heating method, the thermal conductivity and the specific electric resistivity k is given as [7]

$$k\frac{V_h A}{I_h l} = \frac{\left(V_3 - V_1\right)^2}{4\left[2T_2 - \left(T_1 + T_3\right)\right]} \tag{2}$$

where l and A are the length and cross-sectional area of the specimen. I_h and V_h are heating current and voltage drop.

2.4. Pipe method

The pipe method takes advantage of a radial heat flow in a cylindrical specimen. A core heater, which is a tube, rod, or wire, is inserted into the central axis of the pipe-shaped specimen. There are heaters at both ends of the specimen. The combination of the specimen and heaters is surrounded by thermal insulation and then a water jacket or a liquid-cooled heat sink. **Figure 4(b)** shows the schematic and components of the pipe method. End guard heaters can be used to minimize axial heat loss, and also increasing the specimen's ratio of length to

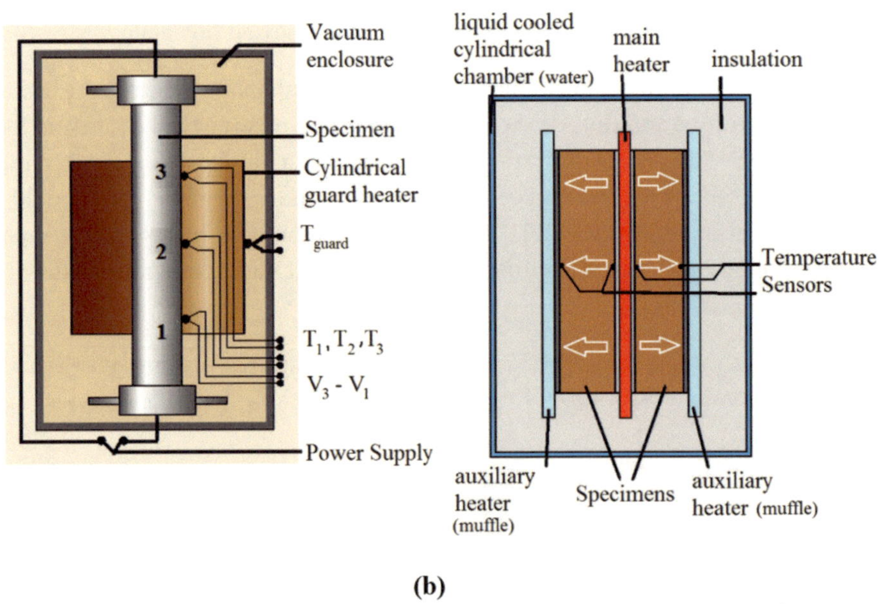

Figure 4. (a) The schematic design of the direct heating method [7], (b) the schematic design of the pipe method.

diameter can achieve the same purpose. The thermal conductivity is obtained by measuring the radial heat flow ϕ [7]:

$$k = \frac{\phi ln(d_2 / d_1)}{2\pi(T_1 - T_2)} \quad (3)$$

This technique has been modified for use with a variety of solids, ranging from insulation materials (20 mW/(m K)) to metals (200 W/(m K)) and evaluating for temperatures between room temperature and 2770 K [7]. This technique can be modified to collect data over time for the simultaneous measurement of thermal conductivity and thermal diffusivity, and these kinds of transient methods are gaining increased interest.

Advantages of the steady-state methods to other methods are as follows:

- simple mathematical expression,
- absolute and primary method for low conductivity specimens,
- acceptable time consumption,
- partially suitable for powdered, granular, or solid forms,
- uncertainties of 1–2% for insulations near room temperatures,
- acceptable small test specimens (except for concentric sphere).

Disadvantages of the steady-state methods to other methods are as follows:

- complexity of the apparatus giving high accuracy,
- uncertainties of 10% or higher to the conditions,
- time consuming,
- immeasurable error due to contact resistance,
- difficulty of measuring geometrically shaped specimens (concentric cylinder or concentric sphere),
- heat losses from especially in parallel plate and concentric cylinder methods,
- difficulty of measurement of heat-flow value for two specimens,
- use error of specimens containing moisture.

3. Transient methods

The advantages of transient methods are mainly distinguished by the short amount time needed, so that various thermal values can be determined in the measurement process.

Therefore, this method is based on a signal measurement and an acceptably small temperature differential. The transient technique is measured by evaluating the feedback response after a signal is transmitted to the specimen for heat generation in the specimen. Therefore, test time is obtained in a few minutes or a subsecond time intervals for transient methods. This method is also more appropriate for high moisture content materials because of the signal and response in the specimen. In many cases, it is possible to replace the temperature measurements at two opposite surfaces with a measurement as a function of time at only one position on the specimen [7].

Among transient methods, the hot-wire and the laser flash methods are commonly used for measuring the thermal conductivity of different materials provided in **Table 1**. A modification of the hot-wire method is the hot-strip or disk technique, which can be applied to solid nonelectrically conducting materials in order to measure the thermal diffusivity and conductivity [12].

3.1. Hot-wire method

This method is a transient technique based on recording the rise in temperature at a defined distance from the heat source. The hot-wire technique is a good method to determine the thermal conductivity of liquids. In the hot-wire method, the specimen preparation is simplified by the use of a heat source, except in the case of solids. When solids are being tested, the wire is situated between two equally sized homogeneous specimens, as shown in **Figure 5**. The hot wire is embedded in small channels, because it is important to ensure sufficiently low contact resistance between solid specimens and the heating wire [7]. For this reason, the hot-wire method is avoided in favor of an increasingly popular variation, the hot-strip method, which is used for measurements on solids.

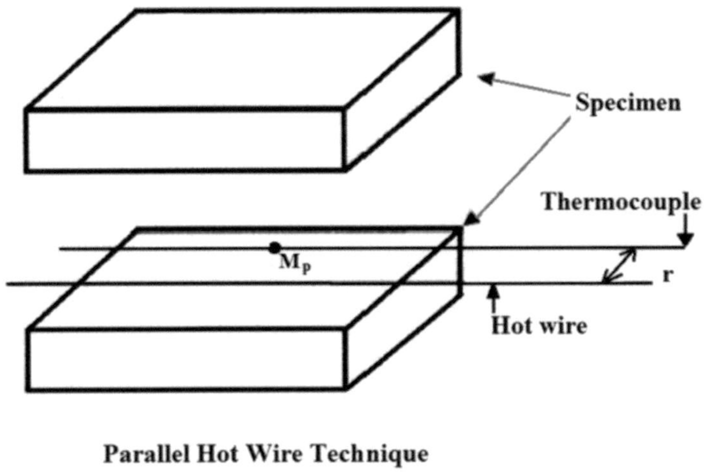

Figure 5. The schematic principle and the design of the hot wire.

In the standard transient hot-wire technique (**Figure 5**), the wire system (often made of platinum) is recorded as two functions of a temperature sensor and heater [7]. One modification of this technique is the probe method, which uses a probe instead of a wire. This probe configuration is useful whenever the specimen conductivity is calculated based on the response of a probe inserted into the specimen. A similar theory underlies both the non-steady-state line heat source (hot wire) method and the probe method for the measurement of thermal conductivity. Both methods are practical for measuring the thermal conductivity of "biological materials, insulations, rocks, ceramics, foods, soils, and glass over a wide range of temperatures" [9]. Thus, the probe method is applied to low-conductivity materials in powder or semirigid form. A closely controlled furnace is used to produce the base temperatures for the tests of the thermal properties of the specimen, soils, in situ [10].

Thermal conductivity is calculated by comparing the plot of the wire temperature versus the logarithm of time, as long as density and capacity are given or measured. The hot-wire method is also capable of measuring the thermal conductivity of gases as well as refractories, such as insulating bricks, fibrous, or powder materials [10]. The technique can be adapted for measuring the properties of liquids and plastics of relatively low thermal conductivity [8]. It is practical for foams, fluids, and melted plastics, but it is impractical for solids [5].

The probe containing the heater and the thermocouple measures instantaneous changes in temperature. Once a predetermined amount of current travels through the heater for a limited period of time, the temperature change of the heater's surface is determined in a characteristic form. After the heat begins to flow from the probe to the specimen side, it reaches the outer surface of the specimen side. When the rise slows down or stops altogether due to heat losses into the environment, the rate of rise with time becomes constant. The thermal conductivity can be calculated based on the linear portion of the temperature versus time curve [8].

The thermal conductivity of the line heat source or probe methods is determined as in Ref. [19].

$$k = \frac{q}{4\pi\left(T(t) - T_0\right)} ln\left(\frac{4t\alpha}{r^2C}\right) \tag{4}$$

neglecting convection and radiation heat losses. Where q is the heat flow per unit length of the source and ln(C) = 0.5772 is Euler's constant. r is the wire radius and α is the sample thermal diffusivity. Thus, the thermal conductivity can be calculated based on the temperature rise at two different times (or the slope of the temperature rise compared to the logarithm of time) and from the strength of the heat source [16].

The use of a stabilized electrical power supply ensures that the heat source produces a constant output. In order to eliminate the interference of axial conduction via the large-diameter current supply, leads are connected to the ends of the hot wire, and also two hot wires of differing lengths are utilized in a differential mode [7]. The other variations of the hot-wire method are the cross wire and the parallel-wire techniques. In the parallel technique, the heater and temperature sensor are separated from each other. In the cross-wire technique, the heating wire is in contact with the thermocouple [20]. The parallel-wire method is advantageous when applied to anisotropic materials and for materials in the magnitude of a thermal conductivity above 2 W/(m K) [20]. The parallel-wire technique can be used for thermal conductivities below 20 W/m K. The cross-wire technique can be used to measure thermal conductivities below 2 W/m K [2].

Another use of the method is a steady-state pipe method having a cylindrical specimen geometry and containing radial heat-flow measurements [7, 21, 22]. This technique is considered to be a transient radial flow technique, so isotropic specimens are required. Although the pipe method has the disadvantage of deviations from the radial symmetric temperature field with respect to the wire method, adequate mathematical model and evaluation procedure can compensate for this disadvantage [7, 23, 24].

3.2. Hot-disk method

The transient plane source (TPS) technique is a recent development of the hot-strip method. It is also known as the Gustafsson probe or the hot-disk method. The technique is designed to measure both thermal conductivity and thermal diffusivity. The advantage of transient technique to steady-state technique is that the effect of the contact resistance is eliminated in the analysis. This method ensures accurate measurements over a thermal conductivity range from 0.005 to 500 W/(m K) over temperatures from 30 to 1200 K [25]. The TPS technique is used for measuring the thermal conductivity of insulation materials and electrically conducting materials [12]. The main advantages of the hot-disk measurement are that it produces results quickly (usually in under 10 min), and that different sensor sizes can be used to accommodate different specimen types. Furthermore, the hot disk requires using specimen sizes that are usually much smaller than those used in other techniques [12].

The hot-strip method is very similar to the hot-wire technique with the exception of an extended strip. The strip is a metal foil between two specimens or a thin metal film on the surface of a plane specimen. Materials studied so far with this system include metals, alloys, ceramics including high conducting ceramics like Aluminium nitride (AlN), high critical temperature materials, minerals, polymers, composites, glass, fabrics, paper, glass wool, foam, powder, biomaterial, and liquids, as well as materials with anisotropic thermal properties [25].

The hot-disk method utilizes a sensor in the shape of a double spiral of nickel covered material. The TPS sensor consists of a number of concentric circles that are made into a double spiral so that the current will travel from one end to the other. A thin polymer coating material is used as electrical insulation and sensor protection on the spiral. The coating materials are most commonly Kapton for measuring temperature ranges between 30 and 450 K, Mica for higher temperatures of up to 1200 K [25], and Teflon. The sensor acts as both a heat source and a thermometer. The source and the thermometer are used to determine the changes in the temperature of the specimen and the increase in the time-dependent temperature, respectively. The sensor is sandwiched between two pieces of the specimen, as shown in **Figure 6**. During testing, a current travels through the nickel spiral and causes an increase in temperature. The generated heat dissipates throughout the specimen on either side. By comparing the temperature versus the time response in the sensor, thermal conductivity or diffusivity can be calculated accurately [26, 27].

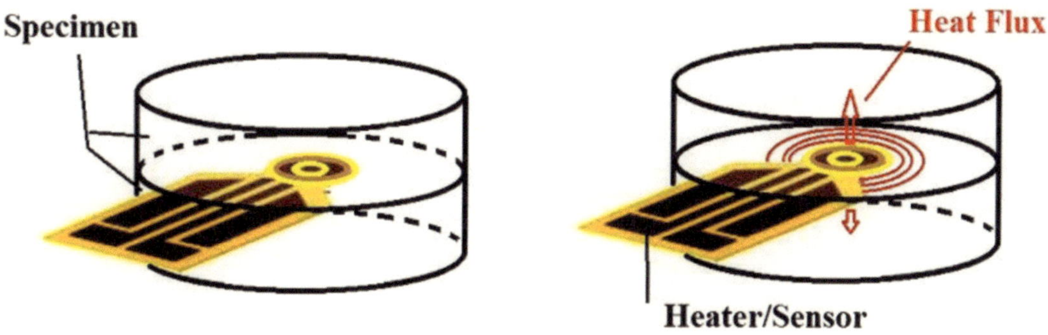

Figure 6. The schematic and principle of the hot disk.

The hot-wire and guarded hot-plate techniques considerably require both larger specimens and a precisely known thickness. While the guarded hot-plate method is a time-consuming method for requiring a temperature gradient across the specimen, the hot-disk method provides instant and direct measurement and the reading is obtained in a short amount of time. In contrast, the guarded hot plate is inherently subject to error due to contact resistance between the thermocouple and the specimen surface. Even at different conductivity levels, a negative influence is unavoidable with the guarded hot-plate method. The hot-disk method, on the other hand, only gathers data regarding from heat diffusion in material for its calculations [26]. As a result, the test procedure of the steady-state method (i.e., the guarded hot plate) is superior to the hot-wire method for testing nonisotropic materials despite the long duration of those steady-state measurements, the higher costs of the steady-state apparatus due to the high reliability in the entire temperature range investigated [28]. It is also in the agreement between the thermal measurements and practical application of insulations [28]. Thus, the hot-wire method should not be employed for determining the thermal conductivity of nonisotropic materials (fiber mats), for which this method is totally ineffective in ranges with a low extinction coefficient, i.e., low bulk densities [28]. On the other hand, one study of the isotropic materials demonstrated that the transient hot-wire method proved to be most dependable due its high reliability and low amount of effort, time, and cost required [28]. For glass fiber, the measured conductivities by hot-disk analyzer are 20 and 12% higher than the claimed values for the lower and higher densities, respectively [12]. For the lower density, the conductivity determined for the rock wool with the hot-disk method harmonizes with the claimed value; the measured value is 8% higher than the claimed conductivity for the higher density [12]. The TPS technique can be used to study liquids, solids, pastes, and powders (electrically conducting or insulating). The TPS technique can provide the values without the interference of thermal contact resistance, without prolonged measurement times, and without extensive specimen preparation [26].

3.3. Laser flash method

The laser flash method is the most commonly used method for ascertaining the thermal properties of solids. The method can investigate to properties of glasses, metals, and ceramics without significant limitations due to uncertainties of the achievable measurement [7]. The property can be measured in a temperature range between −100 and about 3000°C.

In this method, the thermal diffusivity α is determined, and if given the specific heat capacity and the density of a material, the thermal conductivity can be calculated by using the following equation (Eq. (6)) under adiabatic conditions:

$$\alpha = 0.138\frac{d^2}{t_{1/2}} \rightarrow \alpha = {}^{k}\!/_{\varrho c_p} \tag{5}$$

An instantaneous heat pulse is generated by the laser energy. The thermal diffusivity is calculated based on the thickness d of the specimen (typically 2 mm) and the time $t_{1/2}$ [20]. This value represents the time required for the back side surface temperature to reach a value equal to half its maximum value.

In the method, a laser pulse is send to the front side of a specimen, and the temperature change on the back side is measured. The method is conducted through heating a specimen with a short laser pulse of 1 ms width on the front side of the specimen. The temperature increase at its rear side is measured and determined. **Figure 7** shows the schematic and the principle of the method.

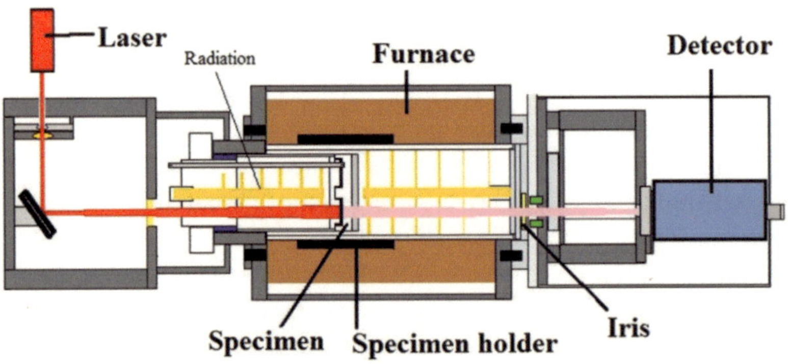

Figure 7. The schematic and principle of the laser flash method [2].

There have been several developments in the method since its introduction by Parker et al. (1961) [29]. The some modifications have been developed to determine directly the thermal conductivity by performing measurement of the specific heat capacity.

The laser flash method has the advantage of involving neither temperature nor heat-flow measurements for the determination of a thermal property. The measurement of the thermal diffusivity is calculated based on the relative temperature change as a function of time only. The main result of this fact is that even at high temperatures the relative measurement uncertainties in the 3–5% range can be achieved [30–32].

3.4. The 3-ω method

The method called the 3-ω method is commonly used for measuring the thermal conductivity of thin films and solid materials. The range of thermal conductivity is changed of 0.20–20 W/(m K), and the range can be extended to 77– 900 K in literature. The method is a similar method to that of the hot wire. While the hot-wire method is a time domain transient technique, the 3-ω method has the advantage of being time independent because it measures electrical signals in a specific frequency domain [33]. An AC current with frequency ω of angular modulation is passed through the wire [34]. The wire is used simultaneously as a heater and a thermometer. The heat generated by this process diffuses into the specimen. Since the electrical resistance of the metal heater is proportional (linear) to the temperature, the temperature oscillation can be measured indirectly by measuring the associated 3ω voltage [34]. Because the current is driven at a frequency ω and the resistance changes at a frequency 2ω, a voltage at 3ω results [34]. In this method, a thin electrically conductive wire is patterned on the specimen, as shown in **Figure 8** [35].

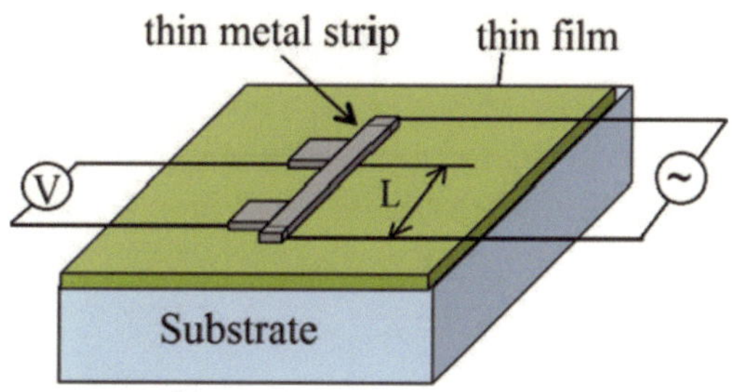

Figure 8. The schematic of the 3-ω method for thin film [35].

Both basic mechanisms generally affect in-plane (x) and cross-plane (z) transport differently, so that the thermal conductivity of related materials is usually anisotropic. The method measures the average thermal conductivity in the in-plane and cross-plane directions. The combination between the heater wire width and the thin film thickness determines the measurement sensitivity according to the in-plane and cross-plane thermal properties of the film [16].

3.5. Fitch method

The Fitch method developed by Fitch is used to measure the materials of low thermal conductivity by using a plane source of heat. This method consists of two components: a heat source and a heat receiver. The heat source is a vessel filled with a constant temperature liquid that functions as a sink. The heat receiver is a sink in the form of a copper plug insulated all sides but the one facing the vessel [36]. The roles of the heat source and the heat receiver can be changed if the vessel is at a temperature lower than that of the copper block. The specimen is interposed between the vessel and the open face of the plug. The sample is firstly in thermal equilibrium with the copper block as shown in **Figure 9**. The vessel is brought into contact with the specimen under a temperature differential. The temperature history of the copper block and the temperature of the bottom of the vessel are measured by thermocouples. It is assumed to have a uniform temperature distribution. A change of time and the temperature are measured, and the thermal conductivity of specimen is calculated using the following equation [37]:

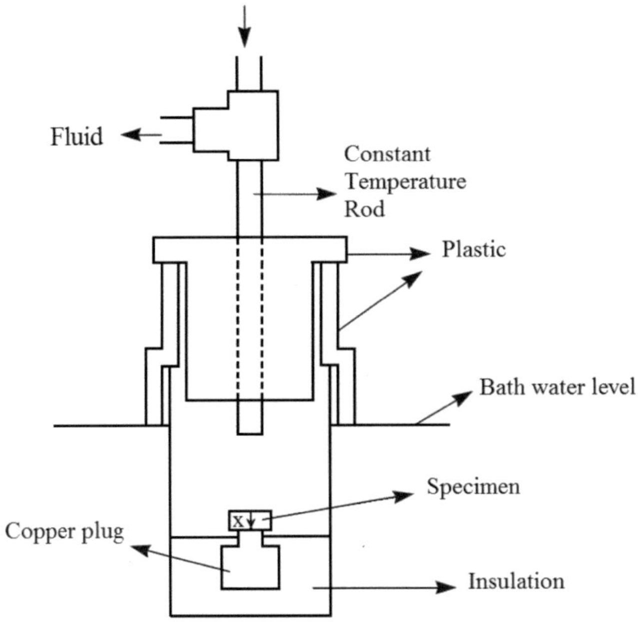

Figure 9. The schematic of the Fitch method [37].

$$k = \frac{\Delta x . m_c . c_{pc}}{A.t} ln \frac{(T_0 - T_\infty)}{(T - T_\infty)} \tag{6}$$

where c_{pc}, Δx, and A are the heat capacity of the copper block, and the thickness and heat transfer area of the specimen. The heat transfer from the copper block to insulation, the heat storage in the specimen, and the surface contact resistance are assumed to be negligible under a linear temperature profile.

This method is generally used for measuring the conductivity of foods and specially small-sized specimens. But the Fitch method is not practical for use at high temperatures. In this method, either a line heat source or one or more heat sources are used [16].

3.6. Photothermal methods

The principle of this method is based on ascertaining the light-induced change in the thermal state of a material in solid, liquid, or gaseous state. After the light is absorbed by a specimen, the changes in temperature, pressure, or density are determined and measured. There are methods in which the specimen is placed in contact with and without the detection system [7].

Measuring the thermal diffusivity measurement via the photothermal method is based on the concept of modulating between the heating of a specimen surface and the temperature phase on the opposite sides of the specimen and includes a frequency function. This technique can be modified by having both surfaces heated simultaneously with a single modulation frequency and taking a measurement of the phase difference between surface signals [20]. The schematic principle of a photoacoustic method is also provided in Reference [38].

4. Standards for thermal insulation measurement

Every measurement of insulation materials has a definition, a standard, and a test realization. The tests are determined by means of experiments for the definition as closely as possible. The experiments are usually performed by standards. When the test realization is obtained, the measurement devices provide its property value. The national and international laboratories use various standards associated with insulation materials. The standards' representation is a main standard to which other representations are compared. The international standards to general methods are listed as follows:

- ASTM C177/C177-13, standard test method for steady-state heat flux measurements and thermal transmission properties by means of the guarded-hot-plate apparatus.
- ASTM C518/C518-10/C518-15, standard test method for steady-state thermal transmission properties by means of the heat-flow meter apparatus.
- ASTM C335/335-10e1, standard test method for steady-state heat transfer properties of pipe insulation.
- ASTM C653-97(2012), standard guide for determination of the thermal resistance of low-density blanket-type mineral fiber insulation.
- ASTM C680-14, standard practice for estimate of the heat gain or loss and the surface temperatures of insulated flat, cylindrical, and spherical systems by use of computer programs.
- ASTM C687-12, standard practice for determination of thermal resistance of loose-fill building insulation.
- ASTM C1303/C1303M-15, standard test method for predicting long-term thermal resistance of closed-cell foam insulation.
- ASTM C1114-06(2013), standard test method for steady-state thermal transmission properties by means of the thin-heater apparatus.

- ASTM C1363/C1363-05/ C1363-11, standard test method for thermal performance of building materials and envelope assemblies by means of a hot box apparatus.
- ASTM C1667-15, standard test method for using heat-flow meter apparatus to measure the center-of-panel thermal transmission properties of vacuum insulation panels.
- ASTM C1696-14ae1, standard guide for industrial thermal insulation systems.
- ASTM C1774-13, standard guide for thermal performance testing of cryogenic insulation systems.
- ASTM D5470-06, standard test method for thermal transmission properties of thermally conductive electrical insulation materials.
- ASTM E1225-09/E1225-13, standard test method for thermal conductivity of solids by means of the guarded-comparative-longitudinal heat-flow technique.
- ASTM E1530-06/E1530-11, standard test method for evaluating the resistance to thermal transmission of materials by the guarded heat-flow meter technique.
- ASTM F433-02 (2009, 2014), standard practice for evaluating thermal conductivity of gasket materials
- ASTM D5334-08, standard test method for determination of thermal conductivity of soil and soft rock by thermal needle probe procedure.
- DIN EN 12667/12939, European standard for measurements of insulating materials using the heat-flow meter method or the guarded hot-plate technique.
- DIN EN 13163, European Standard for characterization of foam insulations for building applications using the heat-flow meter method or the guarded hot-plate technique.
- ISO 8301/8302, thermal insulation—determination of steady-state thermal resistance and related properties—heat-flow meter/guarded hot-plate apparatus.
- ISO 8894-1 (EN 993-14), refractory materials—determination of thermal conductivity—Part 1: hot-wire methods (cross-array and resistance thermometer).
- ISO 8894-2 (EN 993-15), refractory materials—determination of thermal conductivity—Part 2: hot-wire method (parallel)
- ASTM C1113/C1113M-09 (2013), standard test method for thermal conductivity of refractories by hot wire (platinum resistance thermometer technique).

The problem of determining the thermal conductivity of insulation material as a porous medium has been investigated by researchers. The literature contains a large number of standard models for predicting the thermal conductivities of homogeneous, heterogeneous, or composite materials materials [39], as well as methods of a measurement apparatus designed [40]. These models are largely dependent on mathematical models derived from the physical laws that govern the process (e.g., mass and energy balances, thermodynamics, etc.). The literature also present an experimental evaluation of the models and a comparison analysis of the existing experimental methods. Several researchers have proposed a number of models derived from a set of equations, usually based on a basic model, but many of these include empirical parameters [1]. The difference of thermal conductivity values between models and methods can be considered due to the lack of structural data. The structures include nonhomogeneity/homogeneity, the components, the differences in the direction of components, the condensation/evaporation of a liquid phase, the contact mechanism, the size and shape of particles/pores, the tortuosity of pores, mixed phases, shape factor, contact, orientation,thickness, reinforcement etc. [39, 41, 42].

Many thermal conductivity models found in the literature are based on one or more basic structural models, such as the series and parallel models, the geometric mean model, Maxwell models (different forms) and effective medium theory (EMT) models, and the Russell Model [39]. The choice of thermal conductivity models is generally discussed separately in the literature for each of these classes [39]. A detailed description for a bulk material with a porous medium can be expressed in terms of "ancestor," "extra parameter," "component knowledge," and "heat transfer mechanisms" in various models [39]. Similar models are given in different forms and replace the other components instead of a component parameter. It has not been possible to know which expression is the most accurate for describing the thermal conductivity even in every model of two-phase components/mixes. Therefore, there is no single model or prediction procedure to be considered that is universally applicable. In most modeling exercises, the prediction of thermal conductivity also involves a trade-off between simplicity and accuracy. The accuracy of predicted thermal conductivities cannot be reasonably expected to be better than 5% [39–41]. Several models that have been proposed are created specifically for a particular material and contain material-content properties [39].The models need to be revised by clear parameters, particularly with respect to parameters in models (such as thermal conductivities, porosity, or volume fractions ratio of phases). It is concluded that, instead of new relations, the use of present mathematical relations with finite errors for specific materials and at specific intervals is important.

5. Conclusion

The present chapter presents a comprehensive overview of the research progress on the measurement methods of the thermal conductivity of materials. This study discusses the effectiveness of various methods for measuring the conductivity of insulation materials. The models show similarities and differences depending on whether it is a steady state. Some methods are similar or derivative methods regardless of whether the thermal conductivity model being used is similar. The first major conclusion that can be drawn is that these methods may not be applicable for all materials.

Thermal conductivity methods are broadly classified as either steady-state methods or transient methods. In principle, these methods are based on establishing a steady temperature and a signal gradient over a sample. The techniques measure a response as a signal or a temperature given in the sample. Therefore, these techniques are differentiated mainly concerning a range of thermal conductivities, a range of material types, a measurement time, a measurement accuracy, a specimen type, and a temperature range. Estimating low thermal conductivity may be determined with adjustable parameters in these methods.

There are various methods for determining thermal conductivity of insulation materials. The preferable methods are commonly the guarded hot-plate method, the heat-flow method, and then can be given by the hot wire/disk, laser flash methods. Under the condition of anisotropic and isotropic type, some methods are in good agreement with the claimed values. The thermal conductivity value from the method is higher or lower than the value from other methods, depending on the relationships between the heat loss distribution, the surface temperatures, the uniform heat flow over the material, and the signal and its response. However, the methods must be carefully selected according to the materials, such as nonisotropic materials, and the measurement times before applying the methods. It is concluded that choosing methods to use with finite errors for specific structures and at specific intervals is important. This point can provide an effective way and the accuracy and suitability of a variety of methods for users, manufacturers, and researchers in this field.

Many theoretical models, as well as methods, have been proposed by various researchers. The models are based on one or more basic structural models, such as the series and parallel models, the geometric mean model, Maxwell models (different forms), effective medium theory (EMT) models, and the Russell model. The models may not seem to be a clear explanation of the interior structure and behavior of porous materials. The second major conclusion that can be drawn from the studies is that in the future researchers need to standardize methods for insulators or conductors and for thermal conductivity values.

Author details

Numan Yüksel

Address all correspondence to: numan.yuksel@btu.edu.tr; numanyuksel77@hotmail.com

Department of Mechatronics Engineering, Faculty of Natural Sciences Architecture and Engineering, Bursa Technical University, Bursa, Turkey

References

[1] Yüksel N. The investigation of structure and operating parameters effect on the heat transfer coefficient in porous structures [thesis]. Bursa: Uludag University; 2010

[2] Netzsch. E 0214& HFM 436 Lambda thermal insulation materials [Internet]. 2016. Available from: https://www.netzsch-thermal-analysis.com/media/thermal-analysis/brochures/Thermal_Insulation_Materials_E_0214.pdf, https://www.netzsch-thermal-

analysis.com/media/thermal-analysis/brochures/HFM_436_Lambda_en_web.pdf [Accessed: 2016-02-18]

[3] Yüksel N, Avcı A, Kılıç M. The effective thermal conductivity of insulation materials reinforced with aluminium foil at low temperatures. Heat and Mass Transfer. 2012; 48: 1569–1574. DOI: 10.1007/s00231-012-1001-2

[4] Saylor Academy. Thermal conductivity [Internet]. 2016. Available from: http://www.saylor.org/site/wp-content/uploads/2011/04/Thermal_conductivity.pdf [Accessed: 2016-02-18]

[5] K123 of the department of materials engineering and chemistry. Chapter 16- Determination of Thermal Conductivity [Internet]. 2016. Available from: http://tpm.fsv.cvut.cz/student/documents/files/BUM1/Chapter16.pdf [Accessed: 2016-02-18]

[6] Mohesnin N N, editor. Thermal Properties of Food And Agricultural Materials. 1st ed. New York: Gordon and Breach Science Publishers; 1980. 408 p. DOI: 10.2307/2530323

[7] Czichos H, Saito T, Smith L E, editors. Springer Handbook of Materials Measurement Method. 1st ed. New York: Springer Science & Business Media; 2006. 1208 p. DOI: 10.1007/978-3-540-30300-8

[8] TA instruments. The principal methods of thermal conductivity measurement [Internet]. 2016. Available from: http://thermophysical.tainstruments.com/PDF/technotes/TPN-67%20Principal%20Methods%20of%20Thermal%20Conductivity%20Measurement.pdf [Accessed: 2016-02-18]

[9] Evitherm. Measurement methods [Internet]. 2016. Available from: http://www.evitherm.org/default.asp?ID=308 [Accessed: 2016-02-18]

[10] Tong X C. Characterization methodologies of thermal management materials. In: Tong X C, editor. Advanced Materials for Thermal Management of Electronic Packaging. 1st ed. New York: Springer Series in Advanced Microelectronics; 2011. pp. 59–129. DOI: 10.1007/978-1-4419-7759-5

[11] Bankvall C. Guarded hot plate apparatus for the investigation of thermal insulations. Laboratory Methods and Devices. Matériaux et Construction. 1973; 6: 39–47. DOI: 10.1007/BF02474841

[12] Al-Ajlan S A, Measurements of thermal properties of insulation materials by using transient plane source technique. Applied Thermal Engineering 2006; 26: 2184–2191. DOI: 10.1016/j.applthermaleng.2006.04.006

[13] Asdrubali F, Baldinelli G, Bianchi F, Libbra A, Muscio A. Comparative analysis of different methods to evaluate the thermal conductivity of homogenous materials. In: ASME-ATI-UIT 2010 Conference on Thermal and Environmental Issues in Energy Systems; 16–19 May 2010;Sorrento. Italy: CRBnet; 2010. pp. 1–5

[14] Yüksel N, Avci A, Kilic M. The temperature dependence of effective thermal conductivity of the samples of glass wool reinforced with aluminium foil. International

Communications in Heat and Mass Transfer. 2010; 37: 675–680. DOI: 10.1016/j.icheatmasstransfer.2010.01.016

[15] IS 3346, 1980. Method for the determination of thermal conductivity of thermal insulation materials (two slab, Guarded Hot-Plate Method) according to C177 [Internet]. 2016. Available from: https://law.resource.org/pub/in/bis/S02/is.3346.1980.pdf [Accessed: 2016-02-18]

[16] Turgut A. A study on Hot Wire method of measuring thermal conductivity [thesis]. Izmir: Dokuz Eylül University; 2004

[17] BS EN 12664:2001. Thermal performance of building materials and products—Determination of thermal resistance by means of guarded hot plate and heat flow meter methods—Dry and moist products of medium and low thermal resistance. BSI; 2001. 76 p. ISBN: 0 580 36513 1

[18] BS EN 12667:2001. Thermal performance of building materials and products—Determination of thermal resistance by means of guarded hot plate and heat flow meter methods—Products of high and medium thermal resistance. BSI; 2001. 58 p. ISBN:0 580 36512 3

[19] Vargaftik N B, editor. Handbook of Thermal Conductivity of Liquids and Gases. 1st ed. New York, USA: CRC Press; 1993. 398 p. ISBN-13: 978-0849393457

[20] Czichos H, Saito T, Smith L E, editors. Springer Handbook of Metrology and Testing. 2nd ed. New York: Springer-Verlag; Berlin and Heidelberg: GmbH & Co. KG; 2011. 1500 p. DOI 10.1007/978-3-642-16641-9

[21] Assael M J, Dix M, Gialou K, Vozar L, Wakeham W A. Application of the transient hot-wire technique to the measurement of the thermal conductivity of solids. International Journal of Thermophysics. 2002; 23: 615-633. DOI: 10.1023/A:1015494802462

[22] Gustafsson S E, Karawacki E, Khan M N. Transient hot-strip method for simultaneously measuring thermal conductivity and thermal diffusivity of solids and fluids. Journal of Physics D: Applied Physics. 1979; 12: 1411–1421. DOI: 10.1088/0022-3727/12/9/003

[23] Hammerschmidt U, Sabuga W. Transient hot strip (THS) method: uncertainty assessment. International Journal of Thermophysics. 2000; 21: 217–248. DOI:10.1023/A:1006621324390

[24] Watanabe H. Further examination of the transient hot-wire method for the simultaneous measurement of thermal conductivity and thermal diffusivity. Metrologia. 2002; 39: 65–81. DOI: http://dx.doi.org/10.1088/0026-1394/39/1/9

[25] Gustavsson M, Gustavsson J, Gustafsson S, Hälldahl L. Recent developments and applications of the hot disk thermal constants analyser for measuring thermal transport properties of solids. High Temperatures-High Pressures. 2000; 32: 47–51. DOI: 10.1068/htwu259

[26] Hot Disk. Hot Disk/Technology/In-depth [Internet]. 2016. Available from: http://www.hotdiskinstruments.com/technology/in-depth.html [Accessed: 2016-02-18]

[27] Uher C, Morelli D, editors. Thermal Conductivity 25/Thermal Expansion 13. 1st ed. New York: CRC Press; 2000. 391 p. ISBN-13: 978-1566768061

[28] Wulf R, Barth G, Gross U. Intercomparison of insulation thermal conductivities measured by various methods. International Journal Thermophysics. 2007; 28: 1679–1692. DOI: 10.1007/s10765-007-0278-8

[29] Parker W J, Jenkins R J, Butler C P, Abbott G L. Flash method of determining thermal diffusivity, heat capacity, and thermal conductivity. Journal of Applied Physics. 1961; 32: 1679–1684. DOI: http://dx.doi.org/10.1063/1.1728417

[30] Ogawa M, Mukai K, Fukui T, Baba T. The development of a thermal diffusivity reference material using alümina. Measurement Science and Technology. 2001; 12: 2058–2063. DOI: 10.1088/0957-0233/12/12/305

[31] Hay B, Filtz J R, Hameury J, Rongione L. Uncertainty of thermal diffusivity measurements by laser flash method. International Journal of Thermophysics. 2005; 26: 1883–1898. DOI: 10.1007/s10765-005-8603-6

[32] Vozar' L, Hohenauer W. Uncertainty of thermal diffusivity measurements using the laser flash method. International Journal of Thermophysics. 2005; 26: 1899–1915. DOI: 10.1007/s10765-005-8604-5

[33] Choi T Y, Maneshian M H, Kang B, Chang W S, Han C S, Poulikakos D. Measurement of thermal conductivity and convective heat transfer coefficient of water-based single-walled carbon nanotubes solution by modified 3-ω method. Nanotechnology. 2009; 20: 1–6. DOI: 10.1088/0957-4484/20/31/315706

[34] Oh D-W, Jain A, Eaton J K, Goodson K E, Lee J S. Thermal conductivity measurement and sedimentation detection of aluminum oxide nanofluids by using the 3ω method. International Journal of Heat and Fluid Flow. 2008; 29: 1456–1461. DOI: 10.1016/j.ijheatfluidflow.2008.04.007

[35] Gaal D S, Gaal P S, editors. Thermal Conductivity 30/ Thermal Expansion 18. In: Proceedings of the 30th International Thermal Conductivity Conference and the 18th International Thermal Expansion Symposium; 29 August-2 September 2009; Pittsburgh. USA: Destech Pubns Inc; 2010: 1016 p. ISBN: 978-1605950150

[36] Sahin S, Gülüm Sumnu S. Physical Properties of Foods. Food Science Text Series. 1st ed. New York: Springer Science & Business Media; 2006. 251p. DOI:10.1007/0-387-30808-3

[37] Van Gelder M F. A thermistor based method for measurement of thermal conductivity and thermal diffusivity of moist food materials at high temperatures [thesis], Virginia: Virginia Polytechnic Institute and State University; 1998

[38] Georgia Institute of Technology. Heat Lab/Photoacoustic [Internet]. 2016. Available from: http://heat.gatech.edu/photoacoustic/ [Accessed: 2016-02-18]

[39] Yüksel N, Avcı A. The present studies on effective thermal conductivities of porous mediums. Journal of the Faculty of Engineering and Architecture of Gazi University. 2010; 25: 331–346. WOS:000285689400015. ISSN: 1304 - 4915

[40] Eithun C F. Development of a thermal conductivity apparatus: Analysis and design [Thesis]. Trondheim: Norwegian University of Science and Technology; 2012

[41] Yüksel N, Avcı A, The effect of reinforcement on the thermal behavior of insulation materials. Uludag University Journal of the Faculty of Engineering. 2012; 17: 77–90. DOI: 10.17482/uujfe.82424

[42] Gül S H, Karahan M, Yiğit A, Yüksel N. The investigation of thermal conductivity properties of three-dimensional sandwich-woven fabric reinforced composite materials. In: Proceedings of the 14th International Materials Symposium (IMSP'2012); 10–12 October 2012; Denizli. Turkey; 2012. p. 258